The Moralisation of Tourism

Tourism is no longer simply an innocent pleasure, but has been reinterpreted as damaging to cultures and to the environment. 'New' forms of tourism, such as ecotourism, alternative tourism, community tourism and ethical tourism, have been presented as morally superior alternatives to the package holiday. Ironically though, even advocates of the new, ethical tourism brands are increasingly subject to criticisms, not dissimilar to those that they themselves level against package holidays.

The Moralisation of Tourism puts today's critique of tourism in historical context, and outlines a distinctly moral conception of modern tourism, identifying a 'New Moral Tourist', by drawing on a range of contemporary examples. It is a wide-ranging critique, looking at both the effect the New Moral Tourism has on the holidaymaker, and the effect it has on the host societies in terms of development opportunities. Travel that is 'non-intrusive' and 'low impact' is seen as a way of reconnecting with nature and rediscovering lost truths. However, not only does this deference to the destination create new barriers between people – people who are defined by their differences rather than what they have in common – but it also has a negative impact in the field of development.

The Moralisation of Tourism critiques a number of assumptions that are rarely challenged – that the package tourism boom has been destructive and that nature-based tourism is somehow 'ethical'. It argues that eco-friendly and culturally concerned tourism is based on the false premises of environmental and cultural fragility, and of a surfeit of individual freedom for tourists. Using a host of international examples from the industry, the media and non-governmental organisations, Jim Butcher examines what the advocates of 'new tourism' see as being wrong with Mass Tourism, looks critically at the claims made for the new alternatives and makes a case for guilt-free holidays.

Jim Butcher lectures at Canterbury Christ Church University College in Kent.

Contemporary Geographies of Leisure, Tourism and Mobility

Series editor: Professor Michael Hall is Associate Professor at the Centre of Tourism, University of Otago, New Zealand.

The aim of this series is to explore and communicate the intersections and relationships between leisure, tourism and human mobility within the social sciences.

It will incorporate both traditional and new perspectives on leisure and tourism from contemporary geography, e.g. notions of identity, representation and culture, while also providing for perspectives from cognate areas such as anthropology, cultural studies, gastronomy and food studies, marketing, policy studies and political economy, regional and urban planning, and sociology, within the development of an integrated field of leisure and tourism studies.

Also, increasingly, tourism and leisure are regarded as steps in a continuum of human mobility. Inclusion of mobility in the series offers the prospect to examine the relationship between tourism and migration, the sojourner, educational travel, and second home and retirement travel phenomena.

The series comprises two strands:

Contemporary Geographies of Leisure, Tourism and Mobility aims to address the needs of students and academics, and the titles will be published in hardback and paperback. Titles include:

The Moralisation of Tourism
Sun, sand . . . and saving the world?
Jim Butcher

The Ethics of Tourism Development
Mick Smith and Rosaleen Duffy

Tourism in the Caribbean
Trends, Development and Prospects
Edited by David Timothy Duval

Qualitative Research in Tourism
Edited by Jenny Phillimore and Lisa Goodson

Routledge Studies in Contemporary Geographies of Leisure, Tourism and Mobility is a forum for innovative new research intended for research students and academics, and the titles will be available in hardback only. Titles include:

1 Living with Tourism
Negotiating Identities in a Turkish Village
Hazel Tucker

The Moralisation of Tourism

Sun, sand . . . and saving the world?

Jim Butcher

LONDON AND NEW YORK

First published 2003
by Routledge
11 New Fetter Lane, London EC4P 4EE

Simultaneously published in the USA and Canada
by Routledge
29 West 35th Street, New York, NY 10001

Routledge is an imprint of the Taylor & Francis Group

© 2003 Jim Butcher

Typeset in Galliard by
Florence Production Ltd, Stoodleigh, Devon
Printed and bound in Great Britain by
MPG Books Ltd, Bodmin

British Library Cataloguing in Publication Data
A catalogue record for this book is available from the British Library

Library of Congress Cataloging in Publication Data
Butcher, Jim, 1966–
    The moralisation of tourism: sun, sand . . . and saving the world?/
Jim Butcher.
        p. cm.
Includes bibliographical references.
1. Tourism – Social aspects. 2. Tourism – Moral and ethical
aspects.   I. Title.
G155.A1 B89 2002
910′.01–dc21   2002068747

ISBN 0–415–29655–2 (hbk)
ISBN 0–415–29656–0 (pbk)

For Mum, Dad, Jo and the boys

Contents

Acknowledgements

Thanks are due to Dr Richard Sharpley, who made valuable comments on the first draft, and also to the Institute of Ideas think-tank in London whose commitment to critical thinking, has, I hope, rubbed off on this book. Particular thanks are also due to Joanna Williams for critical comments, insights and encouragement as the book came together.

Introduction

I feel rather daunted writing a book about tourism, as it seems that one of the qualifications for writing on the subject is that one must have travelled widely, and become an experienced traveller. I cannot in most cases write from the perspective of 'being there'. My only defence in answer to this criticism is to invoke a well-known saying attributed to the Roman dramatist Terence: 'nothing human is foreign to me'. However, I have never concurred with the critics of tourism who strive to make us feel slightly guilty about our fortnight of fun through their advocacy of 'ethical' alternatives. Theirs is a moralistic agenda of dubious merit to the tourists or their hosts.

Critics of tourism are as old as tourism itself. One hundred and fifty years ago, Thomas Cook was accused of devaluing travel by opening it up to those perceived incapable of cultured behaviour. Whilst the newly ascendant industrial classes looked down worriedly on the drinking and wild behaviour of their workers on holiday, the factory owners themselves were criticised as devaluing the great European cultural capitals as they tried to ape the aristocratic tourists – those considered the masses, and those considered cultured, or individual, has historically been fluid.

Today Mass Tourism is under renewed assault, this time from the advocates of a plethora of types of holiday only united by their antipathy to package tourists. Ecotourism, sustainable tourism, green tourism, alternative tourism and most recently community tourism have been presented as morally superior alternatives to the package holiday. The package holiday revolution, celebrated by some, is increasingly condemned as destructive by a host of campaigns, academics and commentators.

But what is deemed to be so wrong with package tourism? Critics accuse it of environmental degradation. Jonathan Croall's *Preserve or Destroy: Tourism and the Environment* poses the issue in stark terms – either we preserve the environment by reducing the numbers of tourists and adopting ecotourism, or we destroy it.[1] The option of *developing* the environment around needs and wants rarely features in the views of the critics.

Mass package travel is held up as being destructive to culture, too. From the Spanish Costas to the nomadic Masai in Kenya, tourism is held to have destroyed age-old cultures and degraded communities. The tourist is condemned as a harbinger of globalisation, sweeping away diversity in his wake.

In response to the many concerns, tourism has become the subject of a discussion resembling a moral minefield. Where to go, how to act and even whether to go at all, have become subject to a mountain of well-meant advice from self-appointed campaigners, concerned columnists, angst-ridden academics and even marketing gurus eager to amend their products to meet the mood of caution.

Yet the celebrated alternatives, most notably ecotourism, are subject to their own critique. Does it blaze a trail for the masses? Does it expose ever more remote parts of the earth to the threat of tourism? Is it self-defeating – if you are motivated by a belief that tourism is prone to damage cultures and environments, wouldn't you be better off at home? (a conclusion that some have arrived at). Those who do travel are advised to 'travel well' – to seek out and revere the culture of your hosts . . . but not to get too close, for fear of offending cultural sensibilities.

This book sets out to describe and critique a moralistic etiquette surrounding modern leisure travel. The first chapter sets out the *moralisation of tourism* as a contemporary phenomenon and begins to explore some of its characteristics. I argue that leisure travel has been portrayed as essentially environmentally and culturally destructive by a range of people and organisations. Whilst there may not be clear agreement on precisely what is and is not ethical, there is a shared criticism of Mass Tourism, and mass tourists, as exemplary of the destructive nature of economic development. A *New Moral Tourism* is increasingly in evidence, characterised by its advocacy of more 'sensitive' behaviour with regard to environments and cultures. This new school of tourism has acquired a certain sense of moral superiority in relation to its packaged counterpart.

Chapter 2 considers what is really new about the criticisms levelled at package tourists today, given that tourists have had their vocal critics ever since Thomas Cook developed commercial leisure travel for the Victorian working class. I argue that whilst there is some continuity in the low regard for the humble tourist, the way in which they are criticised by advocates of the New Moral Tourism is very different from the past.

Chapter 3 looks critically at the conception of environment that is implicit, and sometimes explicit, in the moralisation of tourism. I argue that there is an assumption that *environment* and *people* exist in an antagonistic relationship to one another, and hence environmental change is routinely interpreted in a one-sided fashion, as destruction. The creative side of development is overlooked.

Chapter 4 addresses the personal freedom long associated with tourism. It is this freedom – the freedom to please oneself – that is questioned by many of tourism's critics. The growth of codes of conduct for travellers and tourists is exemplary of this wariness of personal freedom. I argue that the assumption that cultural encounters are fraught affairs, evidenced in the codes and central to the moralisation of tourism, diminishes the very qualities that make tourism worthwhile in the first place.

In chapter 5 I contend that the moralisation of tourism is a product of disillusionment with modern societies, manifested in a search for the elements New Moral Tourists deem to be missing from their lives. These elements – community, a sense of spirituality and closeness to nature – are located in tourist destinations, most often those interpreted as unmodern, commonly in the Third World. Yet whilst New Moral Tourism is roundly critical of western culture, and both celebratory and precious about the diversity to be found whilst travelling, its own 'enlightened' reading of culture carries conservative assumptions about the societies visited.

In chapter 6 I argue that there is a pervasive sense that tourism contributes to a common global culture, one that is essentially destructive of the societies hosting tourists. Mass Tourism is generally held to be the culprit here. New Moral Tourism, on the other hand, is presented as ethical consumption – an attempt to make a difference to issues held dear through what and where one buys. A range of ethical holidays claim to be 'putting something back', be it through assisting in the conservation of the natural world, or through supporting the way of life of one's hosts. The chapter argues that ethical consumption ends up moralising about exaggerated problems between people, hosts and tourists, and moreover, neglects an assessment of the social inequalities that characterise relationships between nations.

Following on from this, chapter 7 looks at how the moralisation of tourism has important implications for the way development itself is viewed in the Third World. The chapter notes the way that many conservation-oriented non-governmental organisations (NGOs) have adopted ethical brands such as ecotourism as a way to achieve their conservation agenda in the Third World, whilst at the same time claiming to offer innovative development opportunities from nature-based niche markets. I argue that the promotion of such nature-based tourism as an exemplary form of sustainable development reflects profoundly low horizons with regard to the potential to address poverty and inequality. In fact, in basing local economies around becoming guardians of the natural environment, it eschews the transformative economic development that could make a substantial difference to Third World societies.

I offer some closing comments in the postscript.

Having discussed the themes in this book widely over a couple of years, I have found that some people are keen to identify with criticisms of the notion of ethical tourism, but with what I would describe as a fashionable cynicism. 'Gap year travellers just want to mix with the poor, feign concern, and go back home to mummy.' 'Benidorm is more ethical than eco-travel because the resorts help to control the impacts of the tourists by keeping them in one place.' Both of these are comments made to me recently. The book does not go along with the cynicism that sees altruism as 'middle class concern', or holiday resorts as a means of moderating people's effect on the environment. This is not a critique of morality or tourism, but of the *moralisation of tourism*.

Tourists and a Nepalese boy in the Himalayas. The boy
is carrying a traditional basket used to carry agricultural
produce; it is also used by Sherpas assisting trekkers.
(Photo: Karen Thomas)

1 Mass Tourism and the New Moral Tourist

New: 'Markedly different from what was before
Changed, especially for the better
Up-to-date; fashionable'[1]
Moral: 'Concerned with goodness or badness of character'[2]
Mass: 'an aggregate in which individuality is lost'[3]

This chapter identifies and outlines a *New Moral Tourism* – tourism that is justified less in terms of the desires of the consumer and more from the perspective of its perceived benign influence on the natural world and on the culture of the host. This 'tourism with a mission' is explored and contextualised. The chapter gives an overview of the moralisation of tourism, and identifies the main themes of the succeeding chapters.

Mass Tourism – the problem

Modern tourism could be said to have emerged with modern industrial society in the nineteenth century. In this century, industrialisation both spawned the means to travel – initially the railways – and created a growing market amongst the new industrial and professional classes, and amongst the working class, the masses, too. Thomas Cook pioneered leisure travel amongst the middle and working classes in this century. He and his son, John Mason Cook (whose initials JMC are now a brand of Thomas Cook tour operations), took an increasingly broad spectrum of the population to ever more distant destinations. Over the last century and a half the achievement of the industry has been nothing less than the democratisation of leisure travel, from the few deemed worthy, and wealthy enough to partake, to an everyday activity for the majority in developed societies.

The growth of the tourism industry has been driven by economic development. Greater affluence has opened up the possibility to travel for leisure to greater numbers of people. Technical progress – notably the car and air travel – has consistently enabled greater speed, comfort and scope for leisure travellers. Whereas even as recently as forty years ago back-to-back charters were a new innovation, initially confusing to hoteliers and customers, today they are the staple of the big tour operators. The UK's 'big four', Thomas Cook, Airtours, First Choice and Thomson (now part of TUI, the first European-wide package holiday brand, owned by German conglomerate Pressaug) dominate a market that takes annually some thirty-five million British tourists abroad for their holidays. By supplying *en masse*, such companies have lowered the real cost of holidays, and alongside growing incomes, this has contributed to what Vladimir Raitz, founder of Horizon holidays (the first post-war package holiday company to develop charter flight-based packages) refers to as the package holiday revolution. This growth has been mirrored worldwide, with today some 700 million travelling internationally per year for no other reason than leisure. It is estimated that by 2020, there will be some 1.6 billion international tourists.

Flight to the Sun, written by Raitz, and co-authored by travel expert Roger Bray, reflects on the optimism of the post-war boom in tourism. For travel pioneer Raitz, Wordsworth's often quoted lines captured the mood:

> Bliss it was in that dawn to be alive
> But to be young was very heaven.[4]

This optimism was shared by the growing number of customers, for whom a shrinking world represented the opportunity to enjoy snow-capped mountains and sun-soaked beaches.

Tourism has become big business – by some measures the biggest. It employs 74 million people directly, with tourism-related activities estimated to provide some 200 million jobs. It provides the largest source of export earnings for countries as diverse as Spain and Barbados. By 2020 it is predicted that tourism expenditure will top US\$ 2 trillion, or US\$ 5 billion per day. The industry's contribution to global wealth, measured from Gross National Products, is estimated to be 4 per cent directly and 11 per cent including indirect effects.[5] It has also enjoyed consistent growth in recent decades, decades in which some countries have experienced relative decline in some of their traditional industries. Indeed, attracting tourists has increasingly become a preoccupation of politicians and development planners, evidenced by the rise of 'place marketing' and the intense competition to attract sporting and cultural events, World Heritage Status, City of Culture status and a host of other events and designations that can assist in improving a country's share of international tourism receipts.

In economic terms, then, Mass Tourism seems self-evidently vitally important. However, it is increasingly discussed less as an economic phenomenon linked to the creation of jobs and investment, or indeed simply as enjoyment, adventure and innocent fun. Rather tourism has increasingly become discussed as a cultural and environmental phenomenon, and more often than not as fraught and destructive. In this respect the figures for tourism's growth are more likely to be raised in the context of an angst-ridden discussion of its harmful effects than in the celebratory tones characteristic of Thomas Cook 150 years ago, or Vladimir Raitz forty years ago. Wariness rather than celebration typically accompanies accounts of the growth of travel for leisure. It is this emphasis on tourism as a cultural and environmental problem that informs the moralisation of tourism.

This in turn is manifested in a constant denigration of mass package tourism and mass package tourists amongst those for whom such things are deemed unethical. For some, post-war tourism is like Frankenstein's (or perhaps Thomas Cook's) monster, having seemingly run out of control, with dire consequences. The optimism of Raitz, and the association of tourism with innocence, fun and adventure, have been challenged by a mood of pessimism and a sense that moral regulation of pleasure-seeking is necessary in order to preserve environmental and cultural diversity.

The moralisation of tourism involves two mutually reinforcing notions. First, Mass Tourism is deemed to have wrought damage to the environment and to the cultures exposed to it, and hence new types of tourism are proposed that are deemed benign to the environment and benevolent towards other cultures. Second, this ethical tourism is deemed to be better for tourists, too – more enlightening, encouraging respect for other ways of life and a critical reflection on the tourist's own developed society. There are a plethora of terms that academics and those in the industry have applied to this more moral tourism such as ethical tourism, alternative tourism, ecotourism and responsible tourism. Perhaps the term that covers them all, and helps to identify what is distinctive about them taken together, is that coined by industry specialist Ahluwalia Poon – 'New Tourism'.[6] She argues that New Tourism is both an appeal to a certain sense of enlightenment about one's effect on others, and an environmental imperative.

New Tourism – the solution

Poon outlined the marketing aspects of New Tourism thus: the holiday must be flexible and must be able to be purchased at prices that are competitive with mass-produced holidays; holidays are not simply aiming at economies of scale, but will be tailored to individual wants; unlike Mass Tourism, production will be driven by the wants of consumers; mass-marketing is no longer the dominant ethos – holidays will be marketed differentially to different needs, incomes, time constraints and travel interests; the holiday is consumed on a large scale by more experienced travellers, more educated, more destination oriented, more independent, more flexible and more green; consumers of New Tourism consider the environment and culture of the destinations they visit to be a key part of the holiday experience.[7]

Poon clearly considers the New Tourist to be the 'thinking tourist' – more educated, independent of mind and aware. Also, from this definition New Tourism could be regarded as post-fordist tourism – tourism that moves away from a standard, mass-produced product towards a flexible,

individually tailored one, led by individual demands rather than a homogenous mass market.

Poon's identification of post-Fordist production in holidays has resonance.
She quotes the marketing director of British Airways who claims we are
seeing 'the end of mass-marketing in the travel business . . . we are going
to be much more sophisticated in the way we segment our market'.[8] Large
tour operators have adapted accordingly. The big four have bought smaller,
niche operators to tap into the new markets. In addition, despite the squeeze
on so many medium sized tour operators, there has been a large growth in
small, specialised operators, claiming to cater for the specific needs of their
target market. These operators are often keen to identify with a more moral
notion of tourism in their marketing and advertising.

But for Poon, and for many other advocates of New Tourism, it is far
more than dry marketing for 'thinking tourists' – it is an ethical imperative;
it is ethical tourism. As such it is not simply suggested as an option for
prospective tourists, but is advocated as a solution to problems caused by
Mass Tourism. Advocacy, by NGOs, campaigns and New Tourism oriented
tour operators, is a key feature of New Tourism.

For Poon: 'The tourism industry is in crisis [. . .] a crisis of mass tourism
that has brought social, cultural, economic and environmental havoc in its
wake, and it is mass tourism practices that must be radically changed to
bring in the new.'[9] The charge that Mass Tourism has had a generally
destructive impact on host societies is widely asserted in the context of this
advocacy. However, advocates of New Tourism argue that there is a growing
market of more ethical tourists who are rejecting mass-produced, homogenous tourism products in favour of tailored holidays that are kinder to the
environment and benign to the host culture. These people perhaps constitute a new school of 'ethical' tourism – the *New Moral Tourism*. The key
features of their moralised conception of leisure travel are a search for enlightenment in other places, and a desire to preserve these places in the name of
cultural diversity and environmental conservation.

New Moral Tourism – a pervasive agenda

New Moral Tourism is evidenced and expressed in a number of different
types of organisation: governments; companies; and a variety of non-governmental organisations. It is also influential within both popular and academic
discussions of contemporary tourism. As such, it is a pervasive agenda.

The commitment of global government to reforming the tourism industry,
and the tourist, was formalised through the documents that came out of
the 1992 United Nations Earth Summit in Rio. Agenda 21 documentation
for the tourism industry asserts that, 'the travel and tourism industry has a
vested interest in protecting the natural and cultural resources which are the
core of its business'.[10] Elsewhere, the document argues that: 'Travel and
Tourism should assist people in leading healthy and productive lives in

harmony with nature', the industry should 'contribute to the conservation, protection and restoration of the earth's ecosystem', 'environmental protection should constitute an integral part of the tourism development process' and 'tourism development should recognise and support the identity, culture and interests of indigenous peoples'.[11] *Agenda 21 for the Travel and Tourism Industry* also reflects an impulse for education of tourists. It suggests that publicity for the tourist should promote education for ethical tourism, including in-flight videos, magazine articles, and advice on sick bags.

Whilst the efficacy of Agenda 21 is much debated by grass roots environmentalists, this perspective on tourism has been widely taken up by governments and NGOs. Aid agencies around the world have increasingly financed NGOs engaged in ethical tourism development, seeking to generate a rural development sensitive to the natural environment and culture of recipient communities. In the UK the Department for International Development pioneer 'pro-poor' tourism as a means of relieving rural poverty in the Third World. They also support schemes to enlighten prospective tourists, for example, through a recent schools video that portrays package tourists in the most unflattering light.[12] USAID, the aid arm of the United States government, also back up the ethical claims of ecotourism by funding it as a means of generating limited development through ecotourism revenues alongside conservation of the natural environment in the Third World. Promoting an appreciation of the value of conservation for the prospective tourist and their hosts are key aims too.

A host of other quasi-governmental organisations concerned with the environment have also developed a commitment to 'sensitive', sustainable tourism development over the last ten to fifteen years. Their definitions of sustainable tourism are general, but often suggest a preservationist emphasis with regard to the environment and culture. For example, the Federation of Nature and National Parks in Europe, in their influential publication *Loving Them to Death?*, define sustainable tourism as an activity which 'maintains the environmental, social and economic integrity and well-being of natural, built and cultural resources *in perpetuity*' (my italics).[13] This begs the question, central to this critique of tourism's critics, that if they propose to protect nature from the excesses of development, how do they address the poverty and inequality arising from a dearth of development in many parts of the world? Maintaining a society's relationship to its natural environment 'in perpetuity' is hardly likely to tackle this.

Opposition to the perceived excesses of Mass Tourism has been evident in recent years, too, amongst religious and cultural organisations. One event often considered to mark the advent of the global critique of tourism was a conference held in Manila in 1980, convened by a group of religious leaders from developing countries worried about the impact of tourism on local cultures. The 'Manila Statement' boldly asserted that, 'tourism does more harm than good to people and societies in the third world'.[14] The conference also founded the Ecumenical Coalition on Third World Tourism, which has remained highly critical of the tourism industry. A former executive director of the coalition, Koson Srisang, argues that tourism:

> does not benefit the majority of people. Instead it exploits them, pollutes the environment, destroys the ecosystem, bastardises the culture, robs people of their traditional values and ways of life and subjugates women and children in the abject slavery of prostitution . . . [It] epitomises the present unjust world economic order where the few who control wealth and power dictate the terms.[15]

Ecumenical antipathy towards tourism has long been a common theme. The clergy in Britain were vocal in their criticism of the wanton behaviour of early package tourists in the mid-nineteenth century. The Catholic church in Franco's Spain worried about the influence of decadent tourists on Spaniards. Even the Pope recently condemned tourism as 'a kind of subculture that degrades both the tourists and the host community'.[16] However, the criticisms of modern tourism that hold sway are not those seen as conservative and religious, but rather those presented as radical and secular; they are criticisms expressed through a defence of culture and nature. Hence rather than religious organisations, it tends to be conservation NGOs, campaigns, radical academics and journalists who are in the forefront of criticising Mass Tourism and proposing new, 'ethical' alternatives.

There is a diverse range of NGOs involved in the promotion of what they perceive to be ethical tourism. Global conservation NGOs such as the World Wide Fund For Nature (WWF), the Audubon Society and Conservation International increasingly view ecotourism as a means of winning support, both amongst local populations and more widely, for conservation aims. Ecotourism is at the cutting edge of conservation initiatives as it seems to proffer opportunities for people to benefit from preserving their natural environments rather than changing them. Its ethical credentials, then, reside in its ability to combine conservation with limited development goals. More traditional forms of tourism are regarded as less ethical as although they generally yield more in the way of economic development they are deemed to be environmentally destructive and culturally problematic.

More specific projects aimed at particular destinations or types of tourism include Alp Action, the Proyecto Ambiental Tenerife, the Save Goa Campaign and numerous others. In general they highlight the impacts of tourism and lobby against developments they perceive as unethical. The range of goals of these organisations makes any categorisation problematic. However, they often express a disdain for package tourists. For example, the Proyecto Ambiental Tenerife, a project seeking to sustain rural traditions and traditional agriculture on this Spanish island, make the following comment on Mass Tourism:

Mass Tourism was introduced to the island of Tenerife in the 1960s. It made a few local people and many foreigners very rich. It also devastated the rural communities resulting in abandoned terraced farms; beautiful but dilapidated buildings; an age-old culture on the edge of extinction; youth unemployment of 43 per cent.[17]

So whilst the Tenerife and Spanish economies have benefited greatly from tourism, this NGO damns the developments as destructive of tradition. This reverence for tradition over change is characteristic of the moralisation of tourism.

British-based Tourism Concern is prominent amongst the campaigning NGOs. They engage in a wide variety of campaigning activities including lobbying the Gambian government to limit all-inclusive resort developments, lobbying travel companies to pull out of Burma due to human rights abuses there and producing educational materials and codes of conduct encouraging young people to be wary of their impact on the places and peoples they may visit.

In Germany, Studienkreis für Tourismus und Entwicklung (Students for Tourism and Responsibility) operate their prestigious 'To Do!' awards. The winners are almost invariably small scale, locally oriented and green. This organisation, typical of others throughout Europe, state in their aims and objectives that they 'support forms of tourism which contribute to inter-cultural encounter, which allow for joint learning processes, mutual respect as well as respect for cultural diversity and the sustainable use of natural resources'.[18]

In North America, and internationally, The International Ecotourism Society is influential in marketing and promoting the ethical credentials of green holidays. Their role is not just to network with like-minded tourists with a love of the natural world, but to advocate the superiority of eco holidays for both parties concerned: tourists and hosts. The society claim that, 'Ecotravel offers an alternative to many of the negative effects of mass tourism by helping conserve fragile ecosystems, support endangered species and habitats, preserve indigenous cultures and develop sustainable local economies.'[19] They encourage prospective tourists to 'travel with a purpose – a personal purpose and a global one'.

The International Ecotourism Society also work with various development agencies, such as the InterAmerican Development Bank, to advocate ecotourism as an environmentally benign development option. This trajectory looks likely to develop further – it is an aim of the society to develop this, and it also fits in with the 'greening of aid' through nature-based tourism examined in chapter 7.

These and other organisations see raising awareness as a priority. In recent years initiatives with names such as 'Our Holidays, Their Homes', 'Worldwise' and 'Travelling in the Dark' have sought to educate tourists in the UK as to their potential role in environmental and cultural degradation.

Whilst their interest is not restricted to this, there is an emphasis on changing the consumption patterns and the behaviour of holidaymakers in favour of holidays that are deemed benign to the environment and benevolent to the culture of the host. Such organisations have produced ethical codes of conduct, which amount to attempts at a moral regulation of the holiday-maker (examined in chapter 4).

Other NGOs include Kitemark organisations such as the Campaign for Environmentally Responsible Tourism and Green Globe. The former awards their Kitemark to tour operators in the UK they deem to be ethical. Green Globe emerged from the Rio discussions on sustainable development and encourages firms large and small to adapt to the concern over environmental impacts caused by tourists.

Calls for ethical tourism feature ever more prominently in the media, too. Journalist Libby Purves argues that 'Tourists should not travel light on morals', and paints a grim picture of the effects of the industry.[20] The *Guardian* newspaper environment editor, in an article entitled 'Tourism is bad for our health', asserts that Mass Tourism, 'wreak[s] havoc on the environment' and that despite attempts to clean up the industry, 'tourism is essentially and inescapably, environmentally destructive'.[21] Green campaigner and journalist George Monbiot sums up the dim view taken of tourism by media advocates of ethical tourism when he asserts: 'Tourism is, by and large, an unethical activity, which allows us to have fun at everyone else's expense.'[22]

New Moral Tourism is talked up not only as environmentally and cultur-ally benign, as an antidote to Mass Tourism, but also as an 'add-on' to the holiday experience. For example, a new lottery-funded magazine, *Being There*, has recently been launched by British-based campaign Tourism Concern and The Body Shop, aiming to reach 'funky, adventurous, inter-ested and interesting women who want to put something back into the local communities and destinations they visit on holiday'. For the magazine's supporters, travel is a life-changing experience. Anita Roddick argues that the place you visit 'literally goes from being a holiday destination to a place where you can share, learn and grow'.[23] These sentiments are echoed in the web sites of campaigns and the brochures of many nature-based tour operators.

On television, holiday programmes have come in for criticism over their supposed lack of ethical credentials. A recent report castigates British channel ITV's *Wish You Were Here* for not taking sufficient care to encourage thoughtful behaviour on the part of prospective tourists. The compiler of the report argues: 'Editorial content that meets the growing thirst for a rounded insight into a destination will enable viewers to understand the impact their visit may have on the host country.'[25] In this vein, it is not simply tourism itself that is subject to the critical eye of the New Moral Tourism, but also representations of places. These are deemed to appeal to our hedonistic streak, which may preclude ethical consideration. Similar

Holiday snaps – 'The Responsible Traveller', from *Let's Go* guidebooks[24]

'Of course, *Let's Go* readers aren't stereotypical tourists – the purpose of guidebooks like *Let's Go* is to take you off the beaten track and into those places no coach tour would ever dare venture. Unfortunately, where a backpacker leads, the masses are never very far behind. The past decade has seen one hardcore destination such as Thailand "open up" to tourism – and subsequently lose much of their [sic] appeal. And even in places so remote that they are unlikely to ever become major stops on the global trail, insensitive travellers can still have deleterious effects, from the polluting trail of empty coke cans left behind them to offending local people by their unthinking profligacy and disrespect for local customs. Ironically, perhaps, tourists who fly into a resort and don't leave it for the duration of their stay do the least damage – at least the damage has already been done.

'We're not suggesting that you forget a six month trek in the Andes you've been dreaming of for two weeks in Cancun and Marbella – but [there] are some precautions you can take to make sure that your vacation does the least damage to the environment and the indigenous culture as possible.'

points are frequently made with regard to tourist brochures, and even travel guides have been castigated for failing to present what ethical tourism campaigners consider to be an enlightened view. Lonely Planet guidebooks, for example, have recently been subject to a campaign to boycott their Burma guide, on the basis that it encourages travellers to travel to a regime that has used coerced child labour to build up its infrastructure. In fact the guide itself is critical of the regime, too, but takes the view that travellers should decide the ethical issue for themselves. Lonely Planet are also criticised for 'making or breaking' local businesses, depending on whether they are listed in the guides.

These examples are illustrative of the New Moral Tourism. The holiday is re-presented as an arena for ethical behaviour to the benefit of other peoples and the environment, leading to a holiday experience deemed to be far superior. Many of the above assertions present tourists simply as environmental footprints and cultural impositions. That development has a *creative*, as well as destructive, side is rarely alluded to. Indeed, some of the characterisations of modern tourism seem typically to, as one author points out in relation to a different case, modern travel writing, 'attach the word hideous to man-made things, but never to nature'.[26]

Advocacy of New Moral Tourism is also evident in the commercial sector. A host of companies, spurning the four Ss (Sun, Sea, Sand and Sex) in favour of the three Ts (Travelling, Trekking and Trucking) have set out to appeal to the New Moral Tourist. Their advocacy of ethical tourism is often met with scepticism by the NGOs and campaigns, who question whether their concern to be ethical is genuine or merely a marketing ploy. Nonetheless, many such companies echo the criticisms of package tourism made by the NGOs and express a similar commitment to the environment and the host's culture. They also display a similar disdain for package tourists. Explore, a trekking holiday company, have advertised their holidays as being for 'people who want more out of their holiday than buckets of cheap wine and a suntan'. Dragoman view their trucking holidays as visiting places that have been 'shunned by the masses who prefer resorts and beaches'. Other brochures set out the important role of their clientele in relation to supporting the culture and environment of their hosts in the Third World. Encounter Overland regard their customers as 'today's custodians of the ancient relationship between traveller and the native which throughout the world has been the historic basis for peaceful contact'.[27]

Preserving the environment is an important motif of most tours of this type – most donate a small portion of the price paid to organisations engaged in wildlife and environmental preservation. Indeed, the dividing line between private tour operator and conservationist NGO can be a fine one. Discovery Initiatives, for example, works with a number of conservation charities including the World Wide Fund For Nature (WWF), whose Director, Julian Matthews, argues that 'tourism should guarantee that things which draw us now should be the same in 100 years'.[28] Discovery Initiatives donate money to help fund wardens and other resources to help bring about this vision. In similar vein Friends of the Earth have tried to encourage agro-tourism in Cyprus as a counter to the coastal Mass Tourism developments there. Conservation International, a wealthy and influential international conservation NGO, utilises ecotourism as a way to win over local stakeholders to the cause of conservation. They operate their own ecotours to this end. In north-west Bolivia, ecotourists pay large sums to canoe down the Rio Tuichi to stay in stilted cabins on the edge of a lake in the rainforest. Revenue helps to train local inhabitants as guides, cooks and lodge managers, and contributes to Conservation International's goal of rainforest preservation. Such projects clearly involve an orientation towards the eco-consumer, and hence marketing of ecotourism-for-conservation projects is a growing issue for NGOs.

Another example of the link between the conservation NGOs and the commercial world of marketing is a recent venture on the part of Harold Goodwin, well-known British academic and conservation consultant, who founded *responsibletourism.com* as a means of generating markets for ethical, conservation-based tourism products. Many other organisations, such as The International Ecotourism Society and Tourism Concern, operate similar marketing schemes, helping to bridge the gap between conservation organisations and an eco-conscious clientele.

The growing gap year phenomenon is also influenced by the ethical travel imperative. Gap year travel is growing – in 2000, 22,000 British students deferred their university places, and at the time of writing it is estimated that around 40,000 will take a gap year in 2002 (although many do not carry through their gap year plans). Travel visas for Australia – a favourite for gappers – have more than doubled in the last five years.[29] Taking time out to travel is, of course, not new and need not represent anything more than the desire to see a bit of the world. However, the gap year, and young people's travels generally, are increasingly linked to being ethical – doing good for other cultures and for the environment – and a growing number of Gap Year Companies have emerged to provide just this for young (and not so young) idealistic gappers. Gap year travel is increasingly discussed as a passport to a sort of global citizenship and to better career prospects. In this vein the World Expeditions Challenge gap year company quote the Chief Executive of the Universities and Colleges Admissions Service:

> Whatever you might choose to do in your year out, you can be sure you'll not only develop a range of valuable skills, but also have a personally enriching experience, the benefits of which are now widely recognised by universities and colleges.[30]

Another gap year company, Trekforce, organise 'adventure with a purpose' for prospective customers.[31] The projects are focussed on conservation in the Third World, such as rainforest conservation, the construction of a jaguar research centre, work preserving coral reefs in Belize and orang-utan conservation in Borneo. Raleigh International made the news in the UK in 2000 when Prince William took part in a project, which included helping in the building of a wooden cabin in rural Peru.

The much-publicised gap year taken by Prince William and the experience of many others suggest that gap years can be exciting and unique experiences for those inclined to such work. However, the claims to be contributing to these poor societies may be more circumspect. Projects based around preserving the environment are, in truth, unlikely to help in liberating people from poverty. Their ethical credentials seem to come from the *personal* (but very limited) role an individual can play in development, and from a sense of personal mission accompanying such pursuits.

What all the pronouncements from this variety of organisations and individuals point towards is a profoundly negative view of the development of Mass Tourism, and also an appeal, implicit or explicit, for tourists to change their lifestyle and regard their holidays in a different way. It is held that host communities – their environment and culture – and indeed the tourists too, will be the losers if this does not happen. It is suggested that the tourist also benefits from the New Moral Tourism approach by being engaged in something more meaningful and more enlightening than typical package holidays. The influence of these sentiments constitute the moralisation of tourism.

Moral message

Some people have questioned the importance of New Tourism, observing that package holidays remain popular in spite of the assault on their ethical credentials. The extent to which Poon and others identify a sea change in the tourism industry is debatable. The World Tourism Organisation (WTO), picking up on Poon's terminology, estimate that New Tourism will remain below 10 per cent of total tourism for the foreseeable future.[32] In both the developed and developing worlds, New Tourism is peripheral. Also, independent travel and tailor-made tours have always been an option for those who did not want to travel with the package holiday companies (provided, of course, that they could afford it). New Moral Tourism is perhaps not really all that new.

There is, however, evidence of a growth in market segments that we might associate with the moralisation of tourism. According to the World Resources Institute, whilst tourism grew by 4 per cent in the early 1990s, 'nature travel' grew at a rate of 10–30 per cent. World Tourism Organisation estimates show global spending on the more narrowly defined ecotourism market increasing at a rate of 20 per cent per year, about five times the rate for tourism generally.[33]

However, leaving aside the newness of New Moral Tourism in terms of *practice*, it is evident that there is much that is new and changing in terms of the *debates* around tourism. Whilst we may not all be New Moral Tourists, the moralisation of tourism profoundly colours the debates about the future of the industry, and how tourists see themselves. The rise in codes of conduct[34] critical guides promoting ethical tourism (titles such as *The Good Tourist, The Green Travel Guide, Community Tourism Guide* etc.) and the increase in campaign and NGO activity around the issues[35] illustrates that the New Moral Tourism is a prominent moral agenda. The weight given to ecotourism in the burgeoning number of college and university courses featuring tourism, and the talking up of ethical tourism in the media, also points in the same direction.

Holiday snaps – from *The Good Tourist*[36]

'We all joke about going to a Costa, meeting the neighbours, eating fish and chips and drinking English beer, and as this concept becomes more pronounced and the Costas lose their appeal, a new breed of traveller is emerging. Going independent, travelling further in to the interior, choosing somewhere "unspoilt", and demanding more: more ethnic experiences, more genuine culture, more understanding of the people they meet. And they don't want to harm the environment they travel to.'

Even large companies have sought to identify themselves with the environmental and cultural critique of Mass Tourism. For example, British Airways sponsored a recent publication, *The Green Travel Guide*, which was explicitly critical of the growth of tourism – ironically, a growth facilitated by BA, Europe's largest airline.[37] Their advertisement in the guide warns us that, 'It's no use being the world's favourite airline if there's nowhere left worth visiting.' Green campaigners writing in the same publication would undoubtedly blame BA themselves for this state of affairs! STA Travel, a large travel agency catering for the much maligned backpacker and other young travellers, has sponsored a 'Code for Young Travellers' put together by campaigners from Tourism Concern. That a commercial company should be advising their potential customers on what to consume and how to behave is ironic given the dictum 'the customer is always right' – this perhaps should be replaced by '*our* customers are always right' for the purveyors of ethical advice. Both of the examples given here, along with the adoption of ethical environment friendly Kitemarks, and numerous other initiatives, reflect an impulse within the industry to be self-critical and engage with the ethical agenda.

The breadth of deference to the ethical agenda has resulted in an air of moral authority for the New Moral Tourism – it is often simply assumed we must all agree. For example, in *The Green Travel Guide*, Greg Neale, the *Sunday Telegraph* environment correspondent, informs us that:

> Surely we know the damage that modern day mass transport and tourism does: polluted beachlines, once undisturbed hillsides now scarred by the paths of numberless walkers, package holiday jet planes churning out more pollution into the atmosphere, formerly tranquil fishing villages now concrete canyons that reverberate every summer's evening to the beery brayings of tee-shirted tourists.[38]

Presumably, resorts such as Torremolinos come into this category – a place that fifty years ago was a poor, dusty fishing village ('picturesque') and now is a fun-lovers', sun-seekers' mecca ('a monstrosity' in the words of this guide).

A key aspect of New Moral Tourism, then, is *advocacy* – new forms of tourist behaviour (or 'tourism practice') are advocated by a range of public, civil society and commercial organisations with growing influence on the agenda. The advocates have taken the moral high ground. Hence whilst much tourism continues as before, there is a certain etiquette that many are prepared to buy in to – the assumptions implicit in New Moral Tourism are rarely challenged.

Cultural assumptions

Amongst these assumptions is the question of individualism. Poon clearly regards her New Tourist as more 'individual' – less simply 'following the crowd', a view shared by other advocates of New Moral Tourism. Mass

Tourism has long been caricatured as lacking in individualism. The title of one influential book, *The Golden Hordes*, captures the pejorative depiction of package tourists.[39] Another author argues that the growth of alternative tourism is based on a 'search for spontaneity, enhanced interpersonal relations, creativity, authenticity, solidarity and social and ecological harmony', with Mass Tourism seen as running counter to these worthy aims.[40] Poon, who coined the term 'New Tourism', sees Mass Tourism as being 'consumed *en masse* in a similar, robot-like and routine manner, with a lack of consideration for the norms, culture and environment of the host country visited'.[41] These characterisations present holidaymakers as people clearly lacking in the ability to be discerning in what they buy and what they do. New Tourists on the other hand go for more tailored holidays, suited to their own individual needs.

But because many people like a similar environment for their holidays does not make them any less individual, any more than an adventure tourist travelling to a remote Pacific Island becomes a unique individual. New Moral Tourism makes a rather condescending value judgement of how other people choose to spend their money and their leisure time.

New Moral Tourists are presented as being 'people-centred' – interested in the people and the cultures they encounter on their travels. By implication, and often explicitly, Mass Tourists are less people-centred – they are, instead, regarded as 'self-centred', living in a 'tourist bubble'. In this vein prominent Green activist and journalist George Monbiot argues that tourists 'remain firmly behind barriers – be they windows of a coach, the walls of a hotel or the lens of a camera'.[42] Whilst holidays are fleeting visits, and the context of a cash relationship is not always conducive to friendships, one suspects that many tourists who have made friends and mixed easily on holiday would question this.

Compared to the Mass Tourists in their 'tourist bubble', New Moral Tourism is seen as an 'add-on' to the tourist experience. One author says that alternative tourism is tourism that 'sets out to be consistent with natural, social and community values and which allows both host and guest to enjoy positive and worthwhile interaction and shared experiences'.[43] Again, the implication here is that mainstream package holidays are none of these things. The author suggests guided nature walks, bicycle tours, camel safaris, bird safaris and an increase in domestic tourism as worthy alternatives to package tours.

The most well-known marketing typology developed specifically in relation to tourism shares this outlook. Plogg's typology, named after its marketing consultant author, Stanley Plogg, sees tourists as existing along a spectrum, with 'allocentrics' at one end and 'psychocentrics' at the other.[44] Allocentrics are outward-oriented people – interested in people and places. Psychocentrics are concerned with self-gratification – comfort, safety and convenience. It is no surprise that New Moral Tourists are usually seen as Plogg's allocentrics, whilst package tourists are perceived to be psychocentrics.

Whilst Plogg's typology may or may not be a useful device for establishing target markets and selling holidays, his broader assumptions about people are unconvincing. One could argue that New Moral Tourism can reflect a distinct disillusionment with 'people' – family, people at work, people in the neighbourhood and perhaps humanity. After all, is not ecotourism (often at the 'very moral' end of the spectrum) all about eschewing people in favour of a natural high? The New Moral Tourist may be alienated from modern life, seeking respite from 'people' by immersing themselves in nature, or communing with people whose existence is viewed as 'at one with nature'. This response to the pressures of modern life could be regarded as introspective in that it can be accompanied by a self-conscious search for selfhood. The other cultures and environments avidly sought out by 'allo-centric' eco-travellers may comprise a stage for this working out of this modern angst. The Mass Tourist, on the other hand, enjoys conviviality, crowds . . . people.

So which of the two are 'people-centred'? In fact it is possible to reverse some of Plogg's assumptions and arrive at a typology that is at least as convincing as Plogg's own. Whilst the New Moral Tourist may be self-consciously allocentric, perhaps it is the mass package tourist who can lay claim to being more 'people-centred'. The New Moral Tourist, on the other hand, subscribes to the Romantic notion that the self is to be found not in society but in solitudinous contemplation of nature.

Also New Moral Tourists are 'thinking tourists', concerned with the culture and environment of their hosts. Their 'mass' counterparts are cari-catured as unthinking and blind to both the damage they do and the better time they could be having if only they would adopt more ethical practices.

Holiday snaps – what we did on our holidays

For my honeymoon in 1997 we stayed in a flat on Mijas Costa on the Costa del Sol for a fortnight. We had a wonderful time, dividing our holiday between the coastal resorts and towns and villages inland. On returning, a workmate asked me where we had been. 'Southern Spain' I replied. The Costa del Sol sounded a bit common. Whilst Costa del Sol evokes 'crude mass tourism', Southern Spain evokes 'culture'. 'Oh, whereabouts? Did you go to Granada?' Horrified at my lack of cultural capital, I searched for an answer that would keep me in the camp of traveller, and out of that of Mass Tourist. 'Well, we stayed in Mijas – beautiful little place set back from the coast. Lots of tourists, but even more character.' 'Oh how lovely – we've been there, too.' Phew, I thought. My credibility teetered on a knife edge, but I'd come through it. 'We even went to a bullfight . . . errr . . .'. I floun-dered as I realised that for the 'thinking' tourist, bullfights are not 'culture' but barbarism.

One author refers to the way tourists are typically referred to in the third person, and commonly regarded as 'lemmings'. 'We do not know why mass tourists move, but we do know that, at certain times of the year, they all start moving – and we have a fair idea of the destination.'[45] It is in this fashion that the advocates of ethical tourism regard their 'unethical' counterparts – acting as an unthinking mass. But because some people do not engage with the moral tourism agenda, and are not preoccupied with ethical issues related to their consumption of leisure travel, does not make them any less 'thinking'. It may be that they do not consider a holiday as a vehicle for doing good (or bad for that matter).

Whether New Moral Tourism makes us think is debatable anyway. One author suggests that the interpretation of eco-sites should 'seek(s) to reveal meaning and stimulate a cognitive and emotional response. This response should impel people into reconsidering their value base and behaviour.'[46] Eco-holidays and various other niches focussed on 'nature' and 'culture' explicitly share this educative aim. However, the meaning we are to have revealed to us is simply assumed to be the overriding value of the natural environment and the richness of cultural diversity. It is simply assumed that our 'value base' needs shifting in the direction of reverence for our host's way of life. UN advisor Hector Ceballos-Lascurain, often credited as the originator of the term 'ecotourism', echoes this preachy character of New Moral Tourism: 'The person who practices ecotourism will eventually acquire a consciousness that will convert him into someone keenly interested in conservation issues.'[47]

If education is the aim, the focus on the culture of the host society may actually create a barrier. A typical view is that of the Managing Director of travel company Concerning India: 'I do not claim to understand India, only to enjoy and respect its many virtues . . .'.[48] 'Respect' is often invoked in the advocacy of New Moral Tourism to indicate a deference to the culture of the host community. 'They' are deemed so different to 'us' that we cannot know them or make judgements about their society, we can only respect the differences that define us. What we actually learn through this deference is questionable – presumably to claim to be able to understand the history or culture of places visited would run the risk of being accused of cultural arrogance or a lack of 'respect'.

The barrier to education is strengthened by the implicit message of New Moral Tourism to consider one's insignificance in the face of the vast expanse of nature, or the fascinating but bewildering experience of cultures different from one's own. We are encouraged to contemplate the limits of rationality and progress in favour of a celebration of nature and contemplation of spirituality – this is central to the philosophy of ecotourism, the principal moralised brand of leisure travel. There is little room here for critical insight. Even in its own terms of reference – the need to be more informed on our travels about people and places – New Moral Tourism is a stifling etiquette that presents a barrier to discovery.

Holiday snaps – are good causes hijacking holidays?[49]

Two dozen British tourists paid £250 each to take part in Explore Worldwide's Nile Clean-Up Trip in Egypt, picking up dirty toilet paper.

Explore Worldwide brochures promise 'the opportunity to meet ethnic or tribal peoples' (Explore Worldwide).

Discover the World suggests travel can be 'tainted with unease' and it promises packages that can be 'enjoyed with a clear conscience'. (Discover the World).

The stated purpose of the Earthwatch Institute's Amazonian Cultural Traditions volunteering holiday is to record the rich oral traditions of the people of Pirabas 'threatened by the cannonade of modern culture, namely television' (Earthwatch Institute).

Moral or mass?

Many of the cultural assumptions of New Moral Tourism, then, are expressed through distancing these new forms of tourism from mass, package tourism. Responsible tourism, ethical tourism and new tourism – these labels, whilst broad, clearly suggest the previous existence of irresponsible tourism, unethical tourism and old tourism (Mass Tourism), and are attempts to counter these with more moral products. In fact, it may be more useful to consider New Moral Tourism in terms of what it is not, rather than trying to pin down what it is. New Moral Tourism is defined *against* Mass Tourism – according to one author it originated in 'a worldwide reaction against mass tourism'.[50] A slightly less categorical assertion, although expressing a similar sentiment, comes from two prominent authors in the field: 'By the 1990s, there is a sense that the public has become "tired" of the crowds, weary of jetlag, awakened to the evidences of pollution, and in search of something new.'[51]

The stereotypical associations of tourism in its mass form – crude, homogenous, insensitive to hosts, involving resorts that alter the landscape, crowded, frivolous – are railed against by the advocates of a New Moral Tourism.

Hence New Moral Tourism and Mass Tourism can be seen as a series of oppositions. For the New Moral Tourist, Mass Tourism is characterised by:

1 Sameness: It does not involve experiencing cultural differences, being based around a mass-marketed and consumed product in resort complexes, purpose-built for tourists.
2 Crudeness: It involves a lack of self-restraint – alcohol, sex and sunbathing, perhaps in excess.
3 Destructive: Mass Tourism is deemed to be destructive in two senses. It is seen as paying scant regard to the environmental consequences of

tourism. It is also held to involve the imposition of the tourist's culture on to the host, as the former has little interest in the latter. They are there as self-seeking/pleasure-seeking subjects.

In contrast to this New Moral Tourists associate themselves with:

1   Difference: The New Moral Tourist wants to experience cultural and environmental difference and to encourage and sustain that difference. This is done for altruistic motives – Mass Tourism is seen as being bad for the host – but also through a certain deference to the host culture which is held in esteem.
2   Cultural sophistication: The New Moral Tourist takes the trouble to learn about the host's culture and language. Aware of the importance of cultural difference in the host–tourist encounter, the New Moral Tourist adopts a cautious approach, and is sensitive with regard to their behaviour.
3   Constructive: The New Moral Tourist, where possible, will try to be constructive with regard to local cultures and environments. This will involve, for example, buying craft goods from local traders rather than souvenirs (possibly mass-produced, using imported materials) as such goods encourage the preservation of the local culture rather than support a western one. New Moral Tourists may themselves get involved with activities to preserve and sustain a particular way of life, through work on projects, although such assistance may also be in the form of financial support for NGOs and charities, which is sometimes included in the tour cost.

We have, then, two tourism types, the latter opposed to the former:

Mass Tourism	*New Moral Tourism*
Sameness	Difference
Crude	Sensitive
Destructive	Constructive
Modern	Critical of modern 'progress'

These oppositions may be schematic, in that tourism could rarely be characterised as either one or the other. Nevertheless they are the ideological parameters within which tourism discourse, and the self-understanding of the New Moral Tourist, lies. This 'ideal' New Moral Tourist is not a straw man. He encapsulates an important trend that has come to influence how we understand tourism.

The New Moral Tourism defines itself against its Other, Mass Tourism. Here, Mass Tourism is more than a reference to numbers of tourists – it is also, and more crucially, about a *type* of tourist, and a particular type of person. The use of the term 'mass' in the context of Mass Tourism, when

Holiday snaps – Bobos (Bourgeois Bohemians) against the masses[52]

'The code of utilitarian pleasure means we have to evaluate our vacation time by what we have accomplished – what did we learn, what spiritual or emotional breakthroughs were achieved, what new sensations were experienced? And the only way we can award ourselves points is by seeking out the unfamiliar sights, cultivating above-average pleasures. Therefore Bobos go to incredible lengths to distinguish themselves from passive, non-industrious tourists who pile in and out of tour buses at the old warhorse sights. Since the tourists carry cameras, Bobo travellers are embarrassed to. Since tourists sit around the most famous squares, Bobo travellers spend enormous amounts of time at obscure ones watching non tourist oriented pastimes, which usually involve a bunch of old men rolling metal balls.'

not used in a purely descriptive sense, tends to carry pejorative connotations. Mass Tourism is an exemplar of mass consumption in modern, industrial, mass society, and mass consumption is eschewed by the New Moral Tourist.

It is instructive to consider briefly the usage of the term 'mass' in this broader context. Ideas of mass society developed in the latter half of the nineteenth century. They reflected the reality of industrialisation, large conglomerations of people in cities and an attendant fear of the masses – especially when they were organised and politicised. Uses of the term 'masses' at this time carried negative connotations. Generally, the term was used to describe the multitude of 'common' people, perceived as lacking in education, cleanliness and civility. A further association was with disorder – the mass could easily become the rioting mob, acting without recourse to rationality. Finally there was a paternal element to elite conceptions of the masses – they lacked civility, and were therefore in need of civilisation and culture.

These ideas were reflected in the view of early package tourists in Britain and elsewhere. Thomas Cook's first tours were temperance trips, promoting the virtues of abstinence and Godliness. Cook himself held a paternal view of his customers, and he was quite prepared to comment on what he considered their uncouth behaviour. In turn, Cook's critics castigated him for enabling the 'uncultured' masses to partake of leisure travel.

The association of mass with a new type of social form, mass society, was first made by Herbert Blumer in the 1930s.[53] For Blumer, mass society was the object, not the subject of society. The ability of the masses to think critically and act rationally came a poor second to the sense that they were *acted upon*. Mass culture makes us, rather than the other way round, is the logic of this conceptualisation. The masses lack individuality – they are not the rulers of their own destiny, but dupes of voracious advertising. Blumer's view of mass society, whilst it is contested, is important in shaping the post Second World War conception of mass consumerism, and it is a strong undercurrent in the criticisms of modern Mass Tourism.

New Moral Tourism is, then, a crusade against a particular characterisation of Mass Tourism, and the Mass Tourist. Raymond Williams' comment that 'There are no such things as masses, only ways of seeing people as masses' is pertinent.[54] The mass can also be considered the many with a common goal – either threatening or worthy of championing. In the past negative conceptions of the masses would have been contested by political movements and trades unions that stood for the masses, or tempered by a sense that growing affluence for the masses was a sign of progress. Cook himself defended his tours from the critics on this latter basis. However, today, in the absence of a common goal, and without the sense that more opportunities for people to travel is part of human progress, they can be presented as a homogenous, unthinking mass, patronised and talked down to by the self-appointed spokespersons of new, ethical tourism.

Anti-modern morals

As well as a slight on tourists, New Moral Tourism also stands against modernity and transformative economic development. In the view of the New Moral Tourism advocate, for 'transformative', read 'destructive'. The places most often characterised as having been destroyed by Mass Tourism are the Spanish Costas – especially the Costa del Sol. From the Monty Python comedy sketch featuring 'Brits' abroad drinking Watneys Red Barrel and singing 'Torremolinos' to the predictable disparagement from Rough Guide and Lonely Planet guidebook authors, the Costa del Sol has long been stereotypical Mass Tourism.

One author asserts an unequivocal view of Mass Tourism developments such as those in the popular Mediterranean resorts:

> The building of high-rise hotels on beach frontages is an environmental impact of tourism that achieves headline status. This kind of obvious environmental rape is now less common than it was during the rapid growth periods of the 1960s and 1970s.[55]

The high-rise represents mass society – catering for many people, a common standard of accommodation, and the beach front represents a natural encounter between the land and the sea.[56] The latter is sacrosanct for the critics. It is worthy of note that this characterisation of high-rise hotels conveniently located for the beach as 'rape' is not the assertion of environmental campaign literature, but appears in the most widely read textbook on the tourism industry.

Another commentator on the tourism industry makes a similar point in relation to the development of tourism in the Algarve in Portugal:

> This frenzied activity is how it must have been in the South Wales valleys at the start of the industrial revolution: endless digging, building and labouring. In those days the commodities were coal, iron and steel. In the Algarve today they labour for tourism. But the results are similar. Clifftop by clifftop, beach by beach, valley by valley – the natural beauty of the countryside is being eroded. No dark satanic mills or slag heaps, perhaps, but the landscape here is being disfigured just as badly by tower blocks of hotels and apartments.[57]

Not only does this author bemoan 'disfigurement' (like 'rape', implying that nature has human characteristics) of the environment, without any sense that something may be gained, too, from this, but he distances himself from industrial development *per se* on the grounds that it erodes natural beauty. Such emotive assertions, made without qualification, reflect the 'pro-environment–anti-people' character of the critique of 'old' Mass Tourism. And whilst the Costas are iconic of Mass Tourism, package holidays generally are criticised for their effects on the environments and cultures of the destinations.

The critique of modern society implicit in New Moral Tourism is also evident in the idea of a 'post-modern tourist', or 'post-tourist', invoked by John Urry and Maxine Feiffer. The post-tourist reacts against modernism, central to which is 'the view of the public as a homogenous mass'.[58] Urry argues that the weakness of the working class and the growth of the middle class heighten this 'anti-mass' sentiment. One could go further. Regardless of the size of the various social classes, however one might define them, it is the decline of *collectivity*, embodied in political projects of Left and Right, trades unions, church and community that may reinforce the type of individualism exercised by the post-tourist. Moreover, post-modernism's rejection of the idea of progress stands against modernity, and mass consumption in the form of Mass Tourism is exemplary of modernity.[59]

An example of the anti-modern emphasis of New Moral Tourism is the UnTourist network, which appeared in Australia in the 1990s. UnTourists are explicitly seeking out the antithesis of the modern societies:

> we sought out all the most discerning, untouristy people we knew – the insiders, the writers, the foodies, the fishermen, the sailors, the farmers, the culture buffs, the historians and the savvy locals – they helped to hunt out the best of everything the destination could offer in things to do, see, eat, buy, and in places to stay.[60]

This self-conscious search for the 'backstage regions'[61] – those hidden from the less discerning tourist – is characteristic of the New Moral Tourism.

And this is not simply a question of lifestyle – UnTourism is also linked to 'giving something back' to the hosts:

Mass tourism is about infrastructure (big hotels, souvenir shops, garish promotions and the fast buck) whereas Untourism is about caring for people, maintaining unspoiled environments, authenticity and value for money ... If untourists won't go to places created solely to soak up the tourist dollar, preferring to see and do what the locals do ... there will be less room to spoil what is natural, authentic and/or special about a place.[62]

So for this type of tourist, whom sociologist Peter Corrigan argues are becoming more commonplace, taking a stand against modern values (and Mass Tourism) through leisure travel is good for the environment, the communities visited, and, as a more moral form of activity, good for the tourist too.

Sustainable tourism

Much of the scepticism of previous forms of Mass Tourism development is couched in terms of its lack of 'sustainability'. As with so many other phenomena – housing, communities, economy, architecture etc. – tourism has acquired the prefix 'sustainable'. Broadly speaking, certain types of tourism have become strongly associated with being sustainable, and others unsustainable. Typically, ecotourism, nature tourism, green tourism, alternative tourism etc., whilst critically regarded, are placed under the rubric 'sustainable', whilst the package holidays that dominate the market are rarely associated with sustainability. Of course some would argue that Mass Tourism can be sustainable too, and that to focus upon a relatively small section of the tourism market – New Tourism – is to miss the point. But a glance at academic literature, brochures and the literature from relevant NGOs shows that there is a casual association between sustainability and New Moral Tourism brands such as ecotourism that is rarely challenged. This suggests a congruence between the two, at least in the way the term is used.

Sustainable tourism's parental concept, sustainable development, is the aim, apparently, of all manner of organisations, in the commercial world, the public sector and also amongst NGOs. But whilst it is an agenda that is widely bought in to, there is relatively little agreement about precisely what it is. The most common, underpinning, definition of sustainable development is that established in the UN report *Our Common Future* in 1987, and popularised at the UN Earth Summit in Rio in 1992: 'development that meets the needs of the present without compromising the ability of future generations to meet their own needs'. It is generally seen as a response to previous and current forms of growth, deemed to have put the ability of future generations to meet their needs in jeopardy. Development, it is held, has proceeded apace with scant regard for the environment or for its effect on the cultures of the world. There are, it is considered, increasingly pressing environmental and cultural limits to growth.

The association of New Moral Tourism with sustainable tourism, or the 'moralisation' of sustainable tourism, is not surprising. The breadth of usage of its parental concept, sustainable development, seems to have obscured any coherence – the term can be moulded to fit one's preference.[63] Sustainability has always lacked conceptual clarity, and been interpreted in different ways, and may even be seen as inherently contradictory. Its contradictory nature comes out of its attempt to reconcile development and the environment, which for some expresses the problem itself, rather than a solution to a problem. Such a view holds that 'sustainability', at least in the way it is interpreted in the advocacy of New Moral Tourism, and 'development' are in fact mutually contradictory concepts.

One definition of sustainable tourism suggests that the nature–growth contradiction at the heart of sustainable development has been resolved here in terms of nature. The definition given by the Federation of Nature and National Parks in Europe is activity that 'maintains the environmental, social and economic integrity and well-being of natural, built and cultural resources *in perpetuity*'[64] (my italics). Clearly such assertions reflect a preservationist emphasis not just with regard to the natural environment but also to culture. It is an emphasis characteristic of New Moral Tourism.

Indeed, rarely can a term have been so overused as the mantra 'sustainable development' – one source notes that over seventy definitions have been proposed.[65] One definition of sustainable development is that it involves the 'management of air, water, soil, minerals and living species including man, so as to achieve the highest sustainable quality of life'. This amounts to saying that that which is sustainable is, . . . sustainable. It leaves unresolved the question of how we judge this, what our priorities are with regard to the environment and the extent to which we are critical or celebratory about previous patterns of development. Hence without irony, institutions and individuals as diverse as George Bush, the International Monetary Fund, Friends of the Earth and Prince Charles have invoked sustainable development.

From this, it should be no surprise that definitions of sustainable tourism have become numerous, too.[66] It also remains a vague term, one that can be used in a variety of circumstances by a variety of people to convey a variety of meanings. It is a term that can easily be moralised, especially in a climate of cultural uncertainty and environmental angst.

But despite the lack of coherence, it is common to discuss the strength of sustainable tourism as being in its assumption that man is indivisible from the environment.[67] It is viewed as a progressive approach, countering perceived arrogance on the part of humanity in its approach to the natural world. However, this seemingly holistic approach, that views humanity as a part of nature, ignores the reality that all human development has involved a greater ability to harness the natural world for human ends. Upholding this as progressive does not imply ignoring environmental problems, or seeing the environment in purely instrumental terms, but involves a recognition that humanity is a distinctive and dominant part of nature, with the capacity to organise and transform the natural world around human ends.

It is also notable that sustainable tourism has tended to develop increasingly as a *socio-environmental* category, with an emphasis on *people* as well as the effect of development on *ecological processes*.[68] Hence sustainable tourism has developed a profound sensitivity towards cultural change, change in how communities relate to their environments. In the context of tourism, Third World communities are often viewed as guardians of precious environments, and their cultures deemed sustainable on this basis. As I argue in chapter 7, this view has profound implications for how we view the potential to develop poor societies, whose poverty is defined by a reliance on their immediate natural environment.

Ecotourism – an example of New Moral Tourism

As one might expect, given the morally loaded nature of the debate, there is little agreement about precisely what constitutes any of the New Moral Tourism brands. There are many different types of tourism that shelter under the 'ethical' umbrella. Green tourism, ecotourism, alternative tourism, sustainable tourism and community tourism are just a few of these. The plethora of categories confirms that, in reality, there is confusion as to what is and is not new and ethical – what is ethical to one advocate of New Moral Tourism may not be to another.

Ecotourism is strongly associated with being a more ethical form of tourism and is more often than not at the forefront of the moralisation of tourism. The term itself is less than fifteen years old, and since its invention few can agree on a precise definition.[69] For some it is simply tourism to relatively undisturbed areas to appreciate the scenery and wildlife. However, others argue that ecotourism should involve assisting in environmental preservation and developing an environmental conscience – it has a purpose well beyond satisfying the desires of the consumer. A recent dispute over whether fishing could be categorised as ecotourism, revolving around whether it is 'consumptive' or not (the argument was initially around whether the fish should be thrown back, and subsequently over whether they feel pain) is indicative of the nit-picking principles of some zealous ecotourism advocates.[70] More vitally, the originator of the term, Hector Ceballos-Lascurain, saw ecotourism as a means of developing eco-consciousness.[71] It is hence a market segment with a mission – to educate tourists and hosts to lead better lives, more in harmony with nature.[72]

In their 1989 video 'The Environmental Tourist', the Audubon Society, America's foremost environmental advocates, describe ecotourism not as a particular set of practices but as a 'travel ethic'.[73] This chimes with the notion of tourism as having a higher moral purpose, rooted in what is essentially an aspect of lifestyle. This drawing together of lifestyle, and a certain morality linked to the elevation of a certain sense of culture and nature, is central to the moralisation of tourism. It is the common feature that unites the various niches that feature in the search for a new, ethical tourism. They are united in their trepidation at tourism's (and potentially their own) effect on cultural diversity and on the natural environment, and see what we buy and how we behave as a means of exercising more ethical, moral judgement.

One academic paper on the subject of ecotourism to the Ladakh Farms Project in India argues that, for ecotourists,

> travel can mean a lot more than a leisure activity. It might form part of a broader philosophical reflection relating to the self and nature. It might involve trying to find answers to many of the problems experienced when living in a westernised, industrialised country.[74]

The authors quote the following personal communication to illustrate this:

> Many people who have spent time in this ancient culture have found it a life-changing experience. They have come away with a recognition that a life closer to nature is not necessarily one of back-breaking toil. They have been inspired by a new faith in human nature and have often left Ladakh with renewed optimism about the possibility for change in western society.[75]

Visitors to the project, run by the International Society for Ecology and Culture, are encouraged to work and raise the status of subsistence agriculture. According to its advocates, visitors

> have an important role in demystifying the image of the luxury and leisure filled lives that people experience in so-called 'developed' countries. Visitors are expected to educate themselves to educate their hosts through reading *Ancient Futures: Learning from Ladakh* by Helen Norberg Haydn.[76]

Whilst this is perhaps an extreme example of New Moral Tourism, it nonetheless illustrates a number of its important features. First, the rejection of 'western' development and the moral elevation of rural, subsistence, 'sustainable' lifestyles is a common factor. New Moral Tourism is in this sense part of a broader critique of modern society that morally elevates tradition above development as a response to the perceived destructive nature of the latter. However, an irony in the Ladakh Project is that, as the project admits, the crisis arises in part because many younger members of Ladakh society are heading for the towns and cities, presumably less impressed with the benefits of a rural, 'sustainable' life than their western advocates.

Second, the ecotourist seeks enlightenment from the experience. The learning of profound truths, absent in developed societies that are considered superficial, makes ecotourism enriching for the tourist in the view of New Moral Tourism. Whether enlightenment equates to education is doubtful. It would seem that the reverence and 'respect' for tradition provide an obstacle to a critical examination of the grinding poverty of the people of Ladakh.

Third, the Project hopes to enlighten others as to the benefits of a life closer to nature. In this case this applies not only to other prospective tourists, but also to the local people themselves, including those voting with their feet and leaving for the cities. New Moral Tourism, then, is not simply another choice for prospective tourists, but is advocated as a more moral form of behaviour for all of us.

Conclusion

Tourism is becoming increasingly moralised. On the one hand, certain types of tourism, and tourist, are considered unethical, as they fail to recognise a *particular* notion of environmental and cultural risk. On the other, the new, ethical alternatives are seen as not only better from the perspective of the host societies, but also better for the tourists themselves. Consumer choices over what kind of holiday one prefers are transformed into moral choices, seen as having significant consequences for one's host, and also for oneself.

Whilst it is the case that distinct 'new tourism' markets remain relatively small, the moralisation of tourism is a pervasive, fluid agenda, colouring the way we see contemporary leisure travel. It casts a shadow over the growth of leisure travel, a growth that one may have assumed would be viewed in more upbeat fashion. It also questions the notion of innocent fun, traditionally associated with holidays. Simply pleasing oneself has become moral terrain.

A view of the Costa del Sol, taken from the Fuengirola harbour wall. The Spanish Costas are popular with foreign and domestic tourists alike, yet remain a byword for damaging, 'unsustainable' development among many critics. (Photo: Jim Butcher)

2 What's new?

Traveller, tourist and the moral debate

Criticisms of tourism are not new – they have been around as long as people have travelled for leisure. This chapter examines what is distinctive about today's critique of tourism. It argues that 'traveller–tourist', the axis around which snobbery (and the inverted snobbery of those who rail against travellers as 'elitist') has often been expressed, does not fit so well with today's moral distinctions.

From Grand Tours to Cook's tours

Historically, the extension of tourism has always been subject to criticisms – usually from those who wanted to preserve its benefits for themselves. It is easy to see this as a common brand of elitism, with the more privileged trying to differentiate themselves from the uncultured masses. Pronouncements in this vein were commonplace in Victorian Britain, and still today Martin Graham, Chairman of the Federation of Tour Operators, can echo the language of the past, asserting, '[sustainability] hasn't really made any difference to the great unwashed. A lot of people muck up their own back yard, and do just the same on holiday.'[1] One hundred and forty years ago Thomas Cook responded to his critics and defended tourism in a fashion that is also pertinent for today. His customers, he claimed, could be identified by their 'courteous and joyous fraternisation', in contrast to the '"independent" tourists who paid up to three times the cost for the privilege of thus sitting solitary in a crowd of free and elastic spirits'.[2]

On the face of it, it can seem that there is little new in the critique of tourism. Tourists in search of a fortnight of fun and frolics, or just rest and relaxation, remain subject to the critical eye of the concerned. Yet much has changed in the way tourists are derided – there are important aspects of the moralisation of tourism that are distinctly new.

The origin of the word 'tourist' comes from the Grand Tourists. The Grand Tour played the role of a right of passage for the British aristocracy especially in the latter half of the eighteenth century, although Grand Tourists preceded and succeeded this heyday. The Grand Tours were linked to the acquisition of a classical education but represented more than this for the young heirs to estates and fortunes – these travellers were in search of worldliness and a culture that would mark out their right to rule. They presided over the birth of the modern idea of fashion and cuisine as they sampled and brought back ideas from their travels from Paris, Florence, Naples and Venice.

The Grand Tours were no fleeting visits. In fact, the travelling itself made this impossible – often travel, even for the elites, was arduous. Accounts of the Grand Tourists often make reference to the physical fatigue and illness associated with travel at this time. Daniel Boorstin in his seminal essay 'The Lost Art of Travel' quotes an eighteenth-century account thus: 'Under the best conditions six horses were required to drag across country the lumbering coaches of the gentry, and not infrequently the assistance of oxen was required.'[3] The origin of the word travel is 'travail', meaning work, and travelling at this time required great endeavour.

However, alongside the acquisition of culture and the difficult journeys across land and sea, the Grand Tourist exhibited a good deal of the hedonism, licentiousness and recklessness often associated with more mass forms of travel. Young English aristocrats could 'give loose to their propensities to pleasure' according to one observer,[4] and indeed, Casanova's *Memoirs* were an account of his travels through Europe's capitals between 1826 and 1838.

By the end of the Napoleonic wars in 1815 the Grand Tour, previously the preserve of the aristocracy, was increasingly encroached upon by the ascendant industrial elites. The latter provided the means for the extension of travel – the iron from which the railways were constructed, the steam technology that powered the trains – and also comprised a growing class with the means to travel. The tension between town and country, between industry and landed wealth, was to be played out in the cafés of Rome and Naples, where those of landed wealth looked down upon the sons of those whose wealth was gained through industry.

As tourism developed, tourist perceptions changed. Areas of wilderness such as the Lake District and mountainous regions such as the Alps, previously regarded as obstacles to travel, increasingly became the object of travel for travellers influenced by Romanticism. The Romantic spirit of the time is most closely associated with, and best articulated by, the Romantic poets.[5] Prominent amongst these was Wordsworth, who campaigned against the railways coming to his beloved Windermere. Romanticism in essence represented a reaction to modernity – to the growth of cities, urbanisation and industry. It upheld individual emotion against collective experience, and nature against human endeavour. Whilst Romantic travel predates the Industrial Revolution – Rousseau's taste for naturalness over the corruption of development led him to champion the Alps when to many travellers they were seen as an impediment to travel – there is no doubt that industrialisation and the modern ideas of liberty, fraternity and equality, championed by the French Revolution, generated a Romantic reaction that has profoundly coloured tourism in the modern era.

Romanticism not only coloured the tourist's viewpoint, but also shaped the criticisms of the growth of leisure travel. The growing number of tourists exemplified man encroaching on nature, and was therefore undesirable from a Romantic standpoint. Such criticisms were expressed by those who considered leisure their birthright. The extension of travel beyond the aristocracy to the capitalist class was part of the latter's ascendancy. It is hardly surprising, then, that the Grand Tourists should resent this upstart class intruding on their precious Naples and Venice – it signified the end of their dominion. This tension between modern society and the reaction to it has informed the 'traveller–tourist' debate historically.

The Romantic reaction to modernity took in the means of travel as well as the new bourgeois tourists. As early as the 1870s, Ruskin saw the train as devaluing travel, as it 'transmutes a man from a traveller to a living parcel'[6] – sentiments held by many who criticise the package charter holiday industry of today. He also likened it to concentrating one's dinner into a pill. Ruskin's criticism was against bourgeois tourism, but more broadly against the train as a symbol of modern, industrial society. Ruskin rejected the advantages of speed itself: 'All travelling becomes dull in exact proportion to its rapidity.'[7]

Ruskin's reactionary Romanticism was taken to task by Thomas Cook, who celebrated the possibilities opened up by rail travel as: 'Travelling with the millions'.[8] It was the new industrialist tourist who provided much of the market for Thomas Cook's Tours from the 1840s. Cook made travel that bit easier for his clientele by making arrangements for smooth passage, good accommodation and finance (he invented the traveller's cheque) for the growing number of travellers. Cook also pioneered package travel for the working class from the 1850s. Soon large numbers of workers and their families were taking advantage of trips to the seaside – Cook's tours are often regarded as the first mass package holidays.

The growth of leisure travel was bound up with the economic progress wrought by the industrial revolution. The development of industry created both new needs and new possibilities for meeting them. However, in an important sense, tourism was also a spur to industrialisation itself. Tourism brought wealth to coastal areas, which benefited the populations of these areas greatly. Indeed, in certain periods in the nineteenth century, resorts were the fastest growing class of towns. In this sense, tourism was itself part of the industrial revolution, rather than just a product of it.

As well as the development of technology and industry, the new-found mobility for leisure was premised on developments such as holidays with pay for some workers (although this did not become the norm until the 1938 Holidays with Pay Act), the establishment of Wakes Weeks and Bank Holidays as regular holiday times, and growing incomes enabling increasing numbers of workers to engage in tourism. By 1911, 55 per cent of the population of England and Wales were taking day trips to the seaside.[9]

But for the critics of Cook, the new possibilities and new wealth evident in the tourism industry were overshadowed by the notion that this was travel as consumption rather than travel for culture. Novelists, commentators and scholars satirised and scolded the growing number of tourists.[10] The distaste expressed by some at mass, working-class tourism was infused by a broader fear and loathing of the masses that emerged as a reaction to industrialisation, urbanisation, the growth of the organised working class and events such as the Paris Commune that resonated around Europe and struck fear into ruling elites.

The antipathy towards the growth of tourism was directed against the perceived crudeness of these new tourists. They were, as the working class, regarded by many as a race apart from those born into wealth, the latter considered to possess the wherewithal to benefit from travel. On the subject of these early mass tourists the Reverend Francis Kilvert famously commented in his diaries in 1870 that, 'Of all the noxious animals, the most noxious is a tourist; and of all tourists the most vulgar, ill-bred, offensive and loathsome is the British tourist.'[11] Kilvert's comments were unguarded, direct and posited the culture of the masses as inferior to those of his station. In this vein Cook's Tours were criticised for extending travel to the lower orders. For one commentator of the time, cities were

> deluged with droves of these creatures, for they never separate, and you see them forty in number pouring along a street with their director, now in front, now at the rear, circling them like a sheepdog – and really the process is as like herding as may be.[12]

Cook himself had a fatherly approach to his tourists, typical of the outlook of Victorian philanthropists. His first tours for working-class people were temperance trips, where tourists could enjoy a good dose of preaching and bracing walks. An early trip was to the unlikely destination of Leicester, the purpose being a temperance meeting. Bad behaviour was clearly an affront to the missionary in Cook, and on one occasion he was driven to comment that, 'to the shame of some rude folk from Lincolnshire, there have been just causes of complaint at Belvoir Castle: some large parties have behaved indecorously . . . conduct of this sort is abominable, and cannot be too strongly reprobated.'[13]

However, Cook was notable for defending the extension of travel from its detractors. The opportunity to travel for leisure was, for him, part of a more civilised society and the means of travel were a symbol of industrial progress, a progress that was for everyone. In this spirit Cook referred to his tours as 'agencies for the advancement of human progress' as, after all, 'railways and steamboats are the result of the common light of science, and are for the people'.[14] Cook criticised the sheer snobbery of the critics of tourism of his day. For him it was foolish to 'think that places of rare interest should be excluded from the gaze of common people, and be kept only for the interest of the select of society', and he added that, 'it is too late in this day of progress to talk such exclusive nonsense'.[15]

Cook's defence and celebration of progress resonated with the outlook of many people, too. One of Cook's most celebrated ventures, to take ordinary people to London for the Great Exhibition of 1851, attracted vast numbers of customers from all social classes. The Great Exhibition, and other international expositions of the time, were widely fêted as celebrations of scientific endeavour and industry. Cook even wrote a pamphlet on why people should visit the Great Exhibition in which he described it as 'a great School of Science, of Art, of Industry, of Peace and Universal Brotherhood'.[16]

Cook's tours to the countryside also revealed something of the class divisions that were evident in popular tourism. These outings have been described as 'tense occasions' as 'country landowners, however obliging, could not look with complete calm on invading groups of working-class people whisked in by train from nearby towns, for they embodied changes that were only partly understood and were already partly feared'.[17] Hospitality towards one's newly mobile potential adversaries did not come easy.

Having extended the range of his tours from Britain to Europe, and from Europe to Asia and on to Africa, and having established tours within the remit of all social classes, Cook set up a register for people who wanted to be the first tourists to the moon. This was no publicity stunt (people put their names down), rather it reflected a sense that society had, and would continue to, overcome geographical limits to travel.

The first fifty years of the twentieth century were dominated by two world wars. However, tourism developed in this period too. For the working class, improvements in living standards, holidays and transport led to the growth of seaside holidays. The advent of the motor car challenged the pre-eminence of rail travel as the main mode of leisure travel. Car ownership grew to 2 million by 1939, and the car was celebrated as a harbinger of a new freedom for leisure travel.

It is, however, since the Second World War that the modern package holiday abroad has become established. In 1949 Vladimir Raitz audaciously approached the Aviation Ministry to get permission to run chartered flights from Gatwick to Calvi in Corsica, where customers would sleep in tents by the beach (the height of fashion) and drink at a bar brought in specially from Paris. The aircraft used were decommissioned Second World War military aircraft – the war had bequeathed aircraft, airfields and technology that facilitated the rapid growth of international tourism. The modern package holiday was born and grew rapidly as ever-increasing numbers of people were able to travel as tourists. Overseas package holidays developed rapidly throughout the 1960s and 1970s, managing 10 per cent growth per year between 1960 and 1974. Despite a distinct blip in 1974 due to the oil crisis (which put up the price of airline fuel), and the subsequent recession, tourism has continued on an upward trajectory since.

But whilst Raitz regards his industry as ringing progressive changes – 'democratising' foreign leisure travel to increasing numbers of people and boosting the economies of destination areas[18] – it is the boom years of the package holiday industry that are associated with destruction, crudity and ugliness by tourism's contemporary critics.

The results of this boom included Torremolinos on the Costa del Sol, a byword for unethical holidaying or affordable fun, depending on one's view of package travel. For tourism's critics, Torremolinos represents everything that is wrong with the package holiday boom – large scale, cheap, transformative of the environment and of culture. It is roundly criticised even in some of the tourist guides to Spain. For the Lonely Planet guide to *Andalucia* it is 'arguably Europe's finest example of how overdevelopment can ruin spectacular landscapes'.[19] In Cadogan's *Spain*, it is a 'ghastly, hyperactive, unsightly holiday inferno'.[20] The Costas are clearly associated with the excesses – in terms of the physical developments and also the behaviour of some of the visitors – of mass package tourism.

In amongst the carping criticisms one guide points out that on the back of tourism revenue, a considerable amount of money has been spent on the promenade, green spaces and the trawling of thousands of tonnes of sand to build up the beach at Torremolinos, ideal for bathing, families and beach sports. However, such initiatives fail to impress many advocates of New Moral Tourism, relying as they do on development rather than nature. In fact, whilst the development of the Mass Tourism industry has generated a good deal of development, often in poorer regions in Mediterranean countries, the creative aspects of this development are overshadowed by an exaggerated sense of destruction and loss.

Criticisms of the package tourism industry such as these are the starting point of the New Moral Tourism, leading to an impulse to change the industry and the behaviour of its clients. But what is different about the critique of modern Mass Tourism? To what extent is the New Moral Tourism actually new in a qualitative sense? There are two important points to be made here.

First, whilst Cook was part of a tradition that was prepared to defend the extension of travel as an emblem of society's progress, few are prepared to do this today. Cook regarded the extension of tourism to the working class in the following way: 'These are the days of the million . . . [who can] o'erleap the bounds of their own narrow circle, rub off rust and prejudice by contact with others, and expand their sails and invigorate their bodies by an exploration of some of nature's finest scenes.'[21] Very few are prepared to put this case today. Advocates of New Moral Tourism typically argue the opposite, that in fact Mass Tourism is characteristic of mass society, a society in which growth has spiralled out of control, with severe consequences for the cultures and environments of the world. It is in this spirit that 'ethical tourism' dominates discussions, with dissenting voices often adopting a cynical 'we're all doomed anyway so what's the point' posture. In this sense, the discussion of tourism is best understood within the broader assault on the idea of human progress, an idea championed by many 150 years ago, but today often dismissed as at best utopian and at worst fascistic.

Whilst in Cook's era possibilities appeared to be boundless, today there is an acute sense of limits to man's capacity to go further, faster and in greater numbers. The recent grounding of Concorde, an aircraft that had in the 1970s epitomised the spirit of progress in travel, seems to sum up this mood. In fact the means of travel are generally regarded as more problematic than liberating. Notably the car, erstwhile symbol of freedom, has become a symbol of environmental degradation. It is not uncommon for people, including car owners, to see themselves as 'anti-car'. Space tourism may have been a prospect relished by Cook, but today it is regarded with incredulity by many, for whom it seems to epitomise an arrogant humanity, intent on colonising other planets. Only ailing ex-Soviet facilities prospectively offer space tourism, to a clientele of super-rich Americans – a telling symbol of the post Cold War era.

Even seaside piers, which in the past 'involved an attempt to conquer nature, to construct a "man-made" object which at all times and forever would be there dominating either the sea or the sky', are now in disrepair and falling into the sea. Nature, it seems, triumphs over human endeavour. Surviving piers today are championed in the spirit of nostalgia for simpler times past.[22]

Secondly, the debates on tourism in the nineteenth century were to a large extent, demarcated along class lines. The aristocracy were unhappy about the loss of the exclusivity of travel as a new business and middle class emerged. Working-class tourists, the 'great unwashed', were to be avoided both by travelling first class and by travelling to different places. From the 1880s distinctly working class coastal holiday destinations came into their own, and divisions between resorts frequented by the working class, and the gentry and nobility became entrenched, reflecting the class divisions of the time.[23]

Some argue that something similar is true today; that the complaints at the growth of Mass Tourism reflect a desire on the part of the 'new middle classes' to maintain the exclusivity of their holiday pursuits.[24] However, today the critique of tourism is more broadly accepted and often appears not to carry the same snobbishness or elitism that Cook railed against. Package tourists are not so much directly criticised for having inferior moral standards to New Moral Tourists, but are instead criticised for damaging other cultures; the culture of the host populations and hence also cultural diversity. This is different from the past, when tourists were considered to lack the cultural sophistication of the elite. The consequence of this is that today's critics of tourism identify themselves as radicals, champions of 'culture' and 'the environment', unlike in the past when antipathy towards tourism formed part of an unashamed conservative outlook. New Moral Tourism is thus presented as a selfless critique; for others, not just for oneself.

Eminent Victorian gentleman Sir Lesley Stephen spoke of 'Cockney-ridden' resorts, but pointed out that at least the resorts confined the 'swarm of intrusive insects' to one place.[25] The last hundred years have seen the aristocratic sense of superiority decline. Today advocates of sustainable tourism see some value in popular resorts in so far as they keep a lot of people in one place, minimising potential environmental damage. The language of the Victorian critic is unacceptable today, but the sentiment that 'the mass' of people, mobile, constitute a problem is all the stronger for being re-presented in terms of environmental and cultural threats. Whilst Thomas Cook's tours of 130 years ago were criticised for carrying those without the etiquette and cultural understanding of the elite, JMC, the brand used by Thomas Cook package holidays today, is more likely to be criticised for damaging environments and cultures.

From 'traveller–tourist' to New Moral Tourism

The debates over travel in the nineteenth century bequeathed us 'traveller versus tourist', an axis around which the critique of tourism has often been discussed. This is still very much in evidence today, witnessed in tour operator advertising, travel literature and in the broader discussion of the nature of modern tourism. However, as we shall see, travel today is far from immune to the criticisms of its deviant offspring, tourism. Today's critique of tourism, premised as it is on a view of people in opposition to nature and cultures, has extended itself further to encompass the erstwhile critics themselves. The propensity to turn in on itself and to become *self-critical* is an important, new, characteristic of the New Moral Tourism. Today's debate is less 'traveller versus tourist' (although the tourists are the ones generally in the frame), and more a critique that starts out focussing on the package tourist and, following its own logic, ends up criticising the attempts at ethical tourism themselves.

The cultural associations of post-war travel include the Hippy Trail in the 1960s and 1970s. Well-heeled university students, rock stars and assorted hippies established the Trail, stretching from London, picking up in Europe, and on through Asia to Kathmandu. They were self-consciously seeking a spiritual enlightenment, and reacted against what they perceived to be a materialistic, modern society tainted by the Vietnam War. They embodied a rejection, be it a temporary one, of western culture, and placed themselves at the centre of the counter-culture. For the radical thinker Marcuse, Man had become 'one dimensional', and through travel young radicals sought a more human dimension.[26] Travel for them offered a spiritual alternative that had little connection to the growing holiday industry.

The counter-cultural character of the 1960s traveller was epitomised by Jack Kerouac. Kerouac, born into white Middle America, rejected the American dream ticket in favour of a ticket to nowhere in particular – his travel was hedonistic and directionless. Kerouac immersed himself in the culture of oppressed black America – jazz culture. Whilst Kerouac's *On The Road* remains iconic for some travellers, his legacy is unclear. On the one hand, Kerouac's rejection of the American way of life may resonate with New Moral Tourist's moral elevation of the host's culture above their own. On the other, whilst Kerouac is associated with an amoral, directionless detachment from his society, the moralisation of tourism embodies a search for direction and is associated with moral regulation of oneself and others.

Another icon of post-war travel is the Lonely Planet guidebook series. The first Lonely Planet guide was written by Tony and Maureen Wheeler, two enterprising travellers left broke by their travels in the hippy heydays of the early 1970s. It contained useful advice on the merits of Kashmiri and Afghan marijuana – it was not a mainstream publication. Lonely Planet bought into the idea of travel as an alternative to tourism – something more individual, authentic and experimental. Lonely Planet guides today often deride the resorts favoured by package holidaymakers.

Alongside these icons, the symbols of the post-war package tourism industry include Butlin's holiday camps, Horizon Holidays (the first package holiday company to charter aircraft and fly tourists to Majorca and elsewhere) and Blackpool's Golden Mile. Unlike the symbolism of travel, tourism is, for its critics, naff.

So the traveller has generally been presented as a very different animal to the tourist. Travel has long been associated with experience and individualism – travellers are not part of the 'faceless' Mass Tourists, as they are so often portrayed by the ethical lobby. To be a traveller was to take an interest in the local culture – for Tony Wheeler, founder of Lonely Planet guidebooks, to experience it 'at ground level', rather than travel within the 'tourist bubble' of the package holiday industry. American writer Daniel Boorstin set the tone in the 1960s. For Boorstin, 'The traveler was active; he went strenuously in search of people, of adventure, of experience. The tourist is passive; he expects interesting things to happen to him . . . he expects everything to be done to him and for him.'[27]

The self-perception of the traveller is the 'thinking tourist' – someone prepared to strike out, experiment with different ways of life, and not be part of a packaged product put together by global companies. He is someone who takes an interest in the culture and the environment of his host. One could surmise from this that the New Moral Tourist was with us all along, in the guise of the traveller. However, the modern travellers – backpackers, trekkers, ecotourists, gappers – have in many ways increasingly merged into the Mass Tourist – they are increasingly subject to similar criticisms.

The critique of travel is most widely expressed against the burgeoning 'backpacking community'. A report on the behaviour of backpackers paints an unflattering picture – one akin to some of the condemnatory commentaries of package tourists. Dr Heba Aziz of the Roehampton Institute in London draws the conclusion from her research of backpackers in Egypt that they inhabit their own 'international backpacker culture' consisting of sex, drugs, pizzas and pancakes. They hang out with each other in 'den-like coffee shops with thick cushions and hookah pipes' pouring over 'The Book'.[28] 'The Book', as readers of modern backpacker literature such as Alex Garland's *The Beach* and William Sutcliffe's hilarious *Are You Experienced?* will be aware, is the Lonely Planet guide. The conclusion of the research is that travellers, just like tourists, are rather insensitive and crude. And far from breaking away from the tourist bubble to discover other cultures, they are portrayed as happy to create their own culture ably assisted by the Lonely Planet empire.

The tone of this report is echoed in many unflattering commentaries on the backpacker. One recent opinion piece in *The Times* newspaper likened backpackers to 'great lumbering dungbeetles . . . slumping like so many uncollected bin-bags around the Trafalgar Square fountain'.[29] To be young and to travel seems to invite cynicism, if not downright hostility, from some of tourism's critics. The perception of the traveller no longer conforms to that of Boorstin – active, in search of culture and experience – instead they are presented as destructive Mass Tourism in disguise.

Also travellers have often stood some distance from the charge of environmental and cultural degradation, an accusation commonly levelled against Mass Tourism. Today that distance is diminished. One theme that emerges in the new critique of the traveller is that they are at the cutting edge of extending the numbers of people travelling, in a world in which international travel is viewed as already bumping up against limits. Therefore if it is accepted that globally numbers of people travelling is a problem then the travellers inevitably get tarred with the same brush as tourists.

Due to growing incomes and lower-cost travel opportunities, independent travel has grown markedly in recent decades. STA Travel, specialists in independent travel, sent half a million British youth abroad in 1999, many on RTWs (Round The World Tickets) – a tenfold increase on a decade ago.[30] It is estimated that backpacking has increased sixfold over the last ten years, reflecting these trends – upwards of 200,000 people go backpacking annually from the UK alone.[31] Other estimates put travellers at around 10 per cent of international visitors. In particular, lower oil prices and improved technology, coupled with increased competition resulting from airline deregulation, have brought down the cost of travelling by air. In recent years the rapid growth of the low-cost 'budget' airlines such as Easyjet and Ryanair have also improved the opportunities to travel independently in Europe.

Just as important perhaps is what could be regarded as the more flexible existence of young people. Breaks (either voluntary or as a result of short-term contracts or redundancy) in one's working life are more common. Also people are starting families later in life, hence extending the period in which footloose travel is a possibility. Whilst travel has in the past been associated with youth, now the years of youthful freedom have extended into 'middle youth' as more people put off settling down into their thirties and later. Tony Wheeler has pointed out that these days backpackers include everyone from broke teenagers to the 'Hilton Hippies', who travel backpacker style but stay in expensive accommodation; or the 300 doctors at a convention who venture into the Australian outback and stay at a backpackers' lodge.[32] Of course, all this applies to some more than others – career breaks are a pipe dream for those without a career.

The view that this growing band of travellers simply adds to tourist numbers is developed by Brian Wheeller of Birmingham University. For this critic, travel companies promoting ecotourism – definitely in the camp of travel rather than tourism – are simply engaged in a marketing ploy providing a cosy, environmentally friendly feel-good factor for western tourists – meanwhile, global capitalist growth in tourism continues unchecked but in a more acceptable guise.[33] Wheeller argues that, whatever the label, it is all pretty much the same when it comes down to it – it is all pretty negative.

However, the impact of independent travellers is not only seen as a question of numbers – if this were the case the criticisms would be barely distinguishable from those of Mass Tourism. It has been pointed out that backpackers act as *pioneers* of Mass Tourism. In seeking out more remote, less popular locations, the backpacker leaves in their wake a nascent tourism industry. Businesses, local and foreign, spot possibilities to develop the industry, companies cite the location in their brochure and word of mouth increases awareness of the place as a destination. Over time one thing leads to another and a few adventurous trekkers becomes large numbers of youthful pleasure seekers.

One industry expert has noted this role of 'traveller as a destination pioneer' on what was once the Hippy Trail. In a research paper he quotes as evidence Murgha Mack of Top Deck Travel:

> Destinations which were first explored by intrepid backpackers often evolve into mainstream areas for tourism . . . For example, we have carried backpackers overland to Kathmandu for many years, but it is only in the last few years that it has come to be visited by the less adventurous traveller.[34]

And then after the less-adventurous traveller comes . . . the tourist! The travellers spread the tourism net further and further afield as they keep 'a step ahead of the growing army of other travellers and institutionalised mass tourism markets in the continuing search for ever-increasing exotic destinations'.[35] The Hippy Trail, one of our icons of travel mentioned earlier, is now superseded by a trail of 'mass backpackers', who are regarded with the glumness previously reserved for the package tourist.

Travel is also implicated in damaging cultures in a similar fashion to the claims made against Mass Tourism. In Goa and on the eastern coast and islands of Thailand the allegation against independent travel is that it has become commercialised and as such is indistinguishable from mass package travel. This commercialisation, it is argued, has implications for both the host and the tourist.

For the hosts it means they are subject to far-reaching cultural change. The economy becomes geared up to sell to tourists at a profit. An important element of this process is the 'commoditisation' of culture.[36] Here long-standing aspects of indigenous culture become staged as rituals for the travellers/tourists. These rituals may lose their true meaning and value for the community practising them. Their role in cohering the societies is less in evidence as they become bought and sold. Hence travellers – those who look for the traditional, the authentic – may in fact, it is argued, be doing the most harm to the object of their fascination.

For the traveller it is held that the experience of travel is diminished by its commercialisation – a commercialisation that proceeds apace. This is ironic, as travel can be seen as an attempt to escape the commodification of human relationships – the influence of the cash nexus on how we relate to one another. This aspiration, to escape the modern, commercial world, may be confounded, as the attempts themselves are subject to the same commercialisation. This is not a new theme. In the 1960s German writer Hans Magnus Enzenberger argued that, 'Liberation from the Industrial World has become an industry in its own right, the journey from the commodity world has become a commodity.'[37] Rudy Koshar in his book *German Travel Cultures* usefully describes the process of commodification as one of a 'simultaneous multiplication and narrowing of possibilities that late modern and commercial society brings'.[38] More opportunities are available formally, but it becomes increasingly difficult to break out of the particular *way* we consume, relating to each other through the cash nexus. The search for something different – alternative – is seemingly confounded by the commercialisation of travel. Travel becomes commercialised and merges into tourism.

In this vein travel has been problematised by both conservative and radical thinkers. Conservative American critic Daniel Boorstin bemoaned the end of travel in his classic *The Image: a Guide to Pseudo Events in America*.[39] He saw it as denigrated by the decadence of mass production, mass transport and mass communications. People were disinterested, duped, or both. For Boorstin, we are offered, and have become satisfied with, 'pseudo events' rather than the real thing, exemplified by the popularity of 'tourist attractions' over interest in people and places.

Dean MacCannell, writing from a radical perspective, raised a similar theme in his ground-breaking *The Tourist: a New Theory of the Leisure Class* in the early 1970s, written at a time of rapid increase in international Mass Tourism, and also the heyday of the hippy trail.[40] For MacCannell, many people sought out authenticity through their travels. This authenticity was denied to them at home, stifled by the dominance of a society defined by consumerism and the cash nexus. Yet for MacCannell, this search for authenticity is all too often stifled by the commercialised nature of travel and tourism – the search for a more human and humane existence is often in vain. MacCannell hence saw tourism as a metaphor for the contradictions of modern society, a perspective that strongly informs the debate today.

Boorstin and MacCannell are often contrasted to one another – Boorstin dubbed 'elitist' as he seems to view the majority as having been dumbed down to accept 'pseudo' experiences, and MacCannell the radical who portrays tourists as aware of the lack of authenticity in their everyday lives, aspiring to something else; something better. However, the shared theme between these authors is a sense that commercialisation spreads itself ever further and wider, to the detriment of the people concerned, in this case host and tourist. Travel as an attempt to distance oneself from this is either difficult or impossible. Travel is increasingly a part of mass consumer society, something that both authors bemoan. Hence travel is perhaps Mass Tourism – travellers are just fooling themselves.

Our second icon of travel, the Lonely Planet Guide, provides an example of this sense that travel has merged into Mass Tourism. Lonely Planet, once associated with independence, freedom and adventure, is in the frame for spawning generations of 'mass backpackers'. For their critics, these back-packers are really Mass Tourists by another name. In this vein, the Lonely Planet guides have been described as 'a blueprint for the modern package holiday', as 'the book dictates where to stay, where to eat, which day trips to take, what to buy'.[41] In Alex Garland's book *The Beach*, and in the subse-quent film adaptation, the Lonely Planet guide is described as 'The Book', a metaphor often used for backpackers' guides. The Book conjures up the image of the backpacker following religiously the instructions from this 'higher authority' – travelling where The Book recommends, staying where The Book decrees and generally acting as a disciple of the new religion of travel. For critic of travellers Heba Aziz, the guides are deemed to have 'institutionalised' travel, making it similar to, and as damaging as, package tourism.[42]

Today Lonely Planet is a major international operation with millions of sales annually and over 250 publications. And if you haven't got The Book you can access The Web Page from the internet cafés found in every back-packer ghetto. Hence Lonely Planet seems to epitomise the commercialisa-tion of travel. By the critics, its customers are increasingly portrayed less as free-spirited, free-thinking travellers, and more as naive disciples, following The Book religiously, and turning traveller heaven into tourist hell.

The implicit assumption of the assault on travel guidebooks is the same as that underlying criticisms of mass package tourists – that the consumer is an uncritical recipient of falsitudes from a rapacious industry. Travellers, the 'thinking tourists', become part of the faceless, homogenous mass.

Leaving aside the gross cynicism of some of the criticisms of the guides, the charges are surely unjustified. The humble guidebook is just that, a guidebook, with advice that one may choose to follow or ignore. Travellers can take it or leave it. Tony Wheeler, Lonely Planet founder, himself admits to meeting travellers who used his guides to identify places to avoid.

'We're all tourists now'

A backpacking acquaintance of mine sees himself as a traveller rather than a tourist. The only exception, he says, is when he meets other backpackers – then he is only too keen to distance himself from the pretensions of his fellow travellers!

It is common to argue that travelling and tourism are more or less the same thing. However, there are perhaps differences, even if they are more to do with how people like to conceive of themselves than anything else. In general, for travellers the *getting there* is likely to be at least as important as *being there*. Even for the less-experienced, more conservative traveller, travel is associated with an experiential dimension lacking from tourism.

Paul Fussell in his book, *Abroad*, referring to those who see themselves as travellers and look down upon tourists, asserts: 'The anti tourist deludes only himself. We are all tourists now.'[44] This argument that 'we are all tourists now' tends to translate as 'we are all *as bad as* tourists now' cited in the context of a downbeat discussion of the perceived cultural and environmental problems created by tourists and travellers alike. Arch-critic of ecotourism, Brian Wheeller, makes a similar point, arguing that 'new wave tourism' is a sheen on a destructive industry:

> of those considering the issues, popular opinion is that mass tourism with all its stereotyped negative connotations is bad, guilty of causing tourism's impact problems and should be 'dealt with' accordingly.
>
> New Wave Tourism, on the other hand, is seen as the complete antithesis to this, sensitive, sophisticated and sustainable, the perfect antidote to be encouraged and enjoyed. While advocating a more caring industry, it fosters the image of a return (if there ever was one) to the golden age of enriching and educative travel. It echoes Senecas (*sic*) 'we should choose a place which is healthy not only for our bodies but also for our morals.' That this appraisal is perhaps less than objective, coloured (green?) by the fact that many of those voicing an opinion do not regard themselves as mass tourists, or as being part of mass tourism, is completely ignored. For eco tourism, read ego tourism.[45]

Holiday snaps – American writer and traveller Gary Paulsen on travel[43]

'Just once, I went for a run on a Harley motorbike. I thought "Where's the farthest I can go? I'll go to Alaska", It's like 700 miles. A friend of mine, Larry, came along, and if he hadn't, I don't think I would have done it. It rained every day and, God, it was a misery. I realised I had a 26-year-old brain and a 52-year-old butt. But every time I'd look in the mirror, Larry was back there and the two of us pushed each other along. It took us about a month and it poured every day apart from two days when it snowed. We got there and Larry said: "What do you think?" I said: "Hell, I don't know, let's go home . . ." And we turned right around and came back again. I sold the Harley and I don't care if I never see another one.'

Wheeller rightly conveys the sense in which New Moral Tourism is an etiquette that contributes to the moral authority of those adopting it. However, his riposte to 'New Wave Tourism' does not proceed to defend package holidays, but to suggest that ecotourism is just as fraught as its mass counterpart. It is this pessimistic standpoint, rather than the notion itself that we are all tourists, that I would take issue with. Tourism, or travel – whatever your preference – should be celebrated. Tourists and travellers should be defined less by the risks they pose, and more by the opportunities (including purely pleasurable opportunities) they create.

And if there is such a thing as a traveller, as distinct from a tourist, is there anything worth celebrating in this classification, often considered elitist, snobbish and pretentious? In many ways the association of travel with life experience may be just so much hot air. For some, travelling abroad is a break from what they perceive as the routine and stress of modern western existence. If so, the year out could easily be interpreted as simply running away from real life to immerse oneself in a temporary fantasy world for as long as finances permit.

And why cannot 'experience' be gained in other ways? Is the teacher who takes a year out to go trekking and work in Peru more experienced than the one who focuses on her subject for the year and enjoys a fortnight all-inclusive in Spain? Is the student who reads in his bedroom any less wise than the one who volunteers to work in conservation in the Seychelles? And more to the point for many travellers, is working in a bar in Australia so different from working in a bar in London?

I have no reason to defend travellers, never having been part of this fraternity. And it is tempting to go along with the critics who attempt to knock oh so worldly backpackers off their pedestal by rubbishing their experience and pointing out their supposed role in environmental degradation. But the premise of the attack is that people and cultures are fragile, and that we should restrict any desires we have to see the world from a recognition that we are complicit in trashing the planet. This clearly is an attack on all of us, travellers, tourists and everyone else.

New Moral Tourism, then, has tended to broaden the scope of what is to be criticised. The package holiday, Mass Tourism, is the obvious culprit if one starts from the premise that cultural encounters are fraught, that development has spun out of control and that there are simply too many tourists. Backpacking, having expanded rapidly, logically must, and has, entered the frame too.

But what of ecotourism, advocated as a solution to the problems of Mass Tourism? Here is a form of tourism explicitly linked to conservation rather than destruction. Ecotourism, according to the World Tourism Organisation and The International Ecotourism Society in America, is the fastest-growing tourism market (be it from a very low base). So surely, following the logic of the moralisation of tourism, is it not also a risky activity?

In fact, this is precisely the criticism some have made. Ecotourism – for its advocates lying at the opposite end of the moral spectrum to Mass Tourism – is increasingly subject to similar criticisms. One writer, Anita Pleumaron of the Tourism Investigation and Monitoring Team in Thailand, expresses in stark terms the dilemma for the conservation-minded ecotourist. In a paper titled 'Ecotourism or Ecoterrorism' she argues that ecotourism is usually just as destructive as package tourism.[46] Ecotourism, unlike tourism to already-developed regions, threatens 'the expropriation of "virgin" territories'. Moreover, 'travellers have already opened up many new destinations', bringing the Mass Tourism that New Moral Tourism seeks to avoid.

In similar vein, Erlet Cater, advocate of ecotourism, sees dangers in its development: 'There is a real danger that eco tourism may merely replicate the economic, social and physical problems already associated with conventional tourism. The only difference . . . is that previously undeveloped areas are being brought into the locus of international tourism.'[47]

Martha Honey, Director of the Ecotourism Programme at the Institute for Policy Studies in Washington DC, makes a similar point, asserting that

> By definition, ecotourism often involves seeking out the most pristine, uncharted and unpenetrated areas on Earth. Often, these are home to isolated and fragile civilisations. In some areas, ecotourism is at the front line of foreign encroachment and can accelerate the pace of social and environmental degradation and lead to a new form of western penetration and domination of the last remaining 'untouched' parts of the world.[48]

Here, not only is ecotourism complicit in destructive practices, but it is actually 'at the front line of foreign encroachment'. This conjures an image of cultural purity degraded by outsiders. For these authors mass tourists have a single redeeming factor – that they are less interested in 'encroaching' upon wilderness.

The contradictions of ecotourism are also well expressed by a writer in the American Audubon Society magazine: 'Tour boats dump garbage in the waters off Antarctica, shutterbugs harass wildlife in National Parks, hordes of us trample fragile areas. This frenzied activity threatens the viability of natural systems. At times we seem to be loving nature to death.'[49] Such comments are typical of the dilemmas within the advocacy of New Moral Tourism. There is a real irony in all this. Advocates of New Moral Tourism have long criticised package tours as destructive of the environment. Ecotourism was their ethical alternative to the rapacious mainstream industry. Now it stands accused of having the same destructive capacity.

There can be no fixed dividing line between Mass Tourism and New Moral Tourism today, as the latter's criticisms of the former have increasingly turned in on themselves. New Moral Tourism is not just critical of others, but is self-critical too. The premise of the New Moral Tourism –

Holiday snaps – 'Indonesian Tour Guide'[50]

'With the tourists everything is organised, so they don't destroy as much. The traveler wants to see something new and wants it to be cheap and then tells others about it. I prefer tourists . . . the travelers are uncontrolled – they don't want to go to the places already prepared for them, they want to go to other places and then they spoil them – and they don't spend any money.'

that cultural encounters are fraught, that environments are under threat from leisure travel – has meant that increasingly all tourism has come under scrutiny. As one critic puts it, despite ethical intent, 'even small groups of people, or for that matter the lone traveller, no matter how sensitive, may have a disruptive effect on local culture'.[51]

Moreover, for some the discussion has gone full circle. Cynical critics of New Moral Tourism brands argue that as they involve going to remote places and linking with communities, they may in fact be more destructive than Mass Tourism. Such cynics take delight in arguing that perhaps Benidorm and Blackpool are the best examples of environmentally and culturally benign tourism, as the impacts are kept in one place and can be more easily managed. Such arguments portray a dim view of the tourist. These critics clearly regard holidaymakers as willing prisoners in beachside jails. Their critique is a false one – it shares the premises of New Moral Tourism, although at least it is prepared to take it to its logical conclusion.

New Moral Tourism is a fluid, moralistic perspective rather than a defined set of products or activities – although clearly some activities are deemed 'bad' and others broadly 'good'. It is an inclination to see others, and oneself, as potential problems to environments and cultures. It embodies an impulse to lay down a moral code to police the boundaries between tourists and their hosts. Never mind Mass Tourism, travel, for leisure or education, for relaxation or fantasy, is worth defending from this.

Tourists setting out to examine coral on a marine safari in Barbados.
(Photo: Marion Stuart-Hoyle)

3 The host

Fragile places, fragile people?

Two overarching assumptions of New Moral Tourism are that environments and cultures are fragile in the face of growing numbers of tourists. This chapter questions these ideas and suggests that in assuming this, advocates of New Moral Tourism do few favours to the host societies they purport to be sensitive towards.

Natural limits?

The negative views of the tourist referred to in the previous chapters exist alongside and are mediated through a critique of the environmental and cultural problems placed at the door of tourists and the tourism industry. These problems are considered severe, sometimes bordering on apocalyptic. One influential book opens with the assertion: 'A spectre is haunting our planet: the spectre of tourism.'[1] The emphasis on the destructive capacity of tourism is striking. For one critic:

> Over the last 30 years or so, mass tourism has had the effect of ruining landscapes, destroying communities, diverting scarce resources, polluting the air and water, trivialising cultures, creating uniformity, and generally contributing to the increased degradation of life on our planet.[2]

Tourists are commonly charged with being responsible for damaging the environment and destroying cultures. For the critics, the Costa del Sol is no more than a 'concrete jungle' and Goa is rapidly going the same way. It is clearly a serious charge – one that has prompted non-governmental organisations (NGOs), campaigns and aid agencies to devote their energies to reforming the industry, and reforming the tourists themselves. In the former case, new forms of tourism are proposed to replace package holidays, and in the latter case, codes of conduct seek to change the behaviour of the tourist.

Over the last fifty years tourism has increased greatly – there are now around 700 million international tourists annually, compared to 50 million in the mid-1950s. The rapid growth of leisure travel is often presented as being of frightening proportions. There is also an overriding sense that the travel 'carrying capacity' of the planet has been reached or exceeded. Even at the level of the United Nations, some believe society is 'bumping up against limits'.[3] In fact, one could also argue that such a remarkable growth is a drop in the ocean when we consider that it represents approximately one in ten of the world's population.

Whilst there is much concern at the global picture, the issue of tourist numbers is most often discussed in terms of the carrying capacity of a resort, region or village, defined as 'the maximum use of any site without causing negative effects on the resources, reducing visitor satisfaction, or exerting adverse impact upon the society, economy and cultures'.[4] Carrying capacity seems relatively straightforward – the number of people a particular geographical area can cope with, that in turn sets limits to the extent of tourism development.

Yet whilst carrying capacity provides a useful marker for planners, the usage of the term tends to assume a static state of affairs. Carrying capacity is often implicitly assumed as fixed in the New Moral Tourism worldview, determined by the pre-existing relationship between people and the natural environment in a particular place. In fact, development itself, including tourism development, can *transform* the carrying capacity. Better transport, drainage, facilities – for residents as well as tourists – would surely be part of well-planned developments, and would enable greater numbers of people to visit if this was deemed appropriate.

It is ironic that carrying capacity is more likely to be seen as an issue where population density, and density of tourists, is actually low, and conversely is less likely to be problematised in major cities and established resort areas. Concern over capacity is often fraught in rural parts of the developing world, where just a few tourists, it is held, can damage fragile environments and cultures. The problem with this is that it becomes a Catch 22 situation in which development is deemed inappropriate due to carrying capacity constraints, and in which these constraints are never challenged due to the dearth of development.

For example, one study argues that trekkers in the Himalayas are

> pushing against the carrying capacity and hence the sustainability of the regional environment. The wood demanded and the biodegradable litter created by the several hundreds in the 1950s were sustainable within the system's productivity, but the demands and refuse of the many thousands in the 1900s are not.[5]

Here the carrying capacity is determined by the pre-existing relationship between people and the environment in this area. There is no suggestion that greater numbers of tourists could generate revenue that may have the effect of altering this relationship.

Some would argue, though, that the carrying capacity, if exceeded, will put off ecotourists, trekkers and green tourists and damage the economy – that in a sense attempts to expand may be self-defeating. This is, however, a wholly different argument, which amounts to saying that the market won't stand for environmental change. It may be entirely rational to place a carrying capacity on an area of land to this end. However, it has little to do with fragility.

The notion of carrying capacity is sometimes modified to allow for 'Limits of Acceptable Change', or 'adaptancy', referring to the limits to changing carrying capacity progressively over time. Such an approach is more sophisticated, and undoubtedly useful in planning. However, it cannot answer the question as to what these limits are – for the advocates of New Moral Tourism, environmental and cultural imperatives rein in substantial change by branding it as destructive. LAC, and the notion of adaptancy, also present change as an exogenous variable to the societies in question. It can easily be interpreted in a mechanical fashion that denies the creativity of the host societies.

Fragile environments?

Of course, some areas may be regarded as being more 'fragile' than others – there is an argument that different areas, defined by their culture as well as geography, have a differential ability to cope with development. Two authorities define fragile environments as 'environments that are less suitable for dense human colonisation'[6] and point out that such areas may be attractive as potential tourism destinations due to their natural beauty. But in what sense are these areas less suitable for human habitation? Industrial development, cities, developed infrastructure and other features of modern societies exist in many different climates and geographical conditions, from London to Rio de Janeiro, and from Seoul to Addis Ababa. The reasons for development in some parts of the planet and for a lack of development in others are primarily social rather than the result of natural suitability or fragility. Indeed, if natural suitability were a criteria, California would never have been built, and rebuilt, on the San Andreas fault line, parts of Holland would be in the sea and Hackney Marshes would still be a marsh, rather than a residential area in London.

Two authorities, Harrison and Price, see fragile environments as including those that exhibit 'marked seasonality, which means that many human activities are limited to quite clearly defined parts of the year'. Such activities are listed as 'cultivating crops, collecting naturally growing foods, hunting, or fishing [which are] typically limited to relatively few months – or even weeks – of the year'.[7] These months or weeks may also be the seasons for tourism, hence the propensity for tourism to damage such fragile environments. Yet societies that rely on the land in this way are often less developed, exhibiting high infant mortality and low literacy levels.

The dilemma here is that if the problematic consequences of incremental development are seen as a reason to question the appropriateness of any development, then the environment remains fragile. It becomes a self-reinforcing process. The authors are correct to point out that 'Tourism should not be viewed in isolation from the environments and cultures on which it is imposed'[8] but similarly, those environments and cultures do not exist outside of an existing level of development and can be understood only in their relationship to this, not as prior features of a society. Posed in this way, it is possible to view environmental capacity problems less as too much development and more as the partial nature of development (lack of infrastructure alongside resort developments) *or even the overall lack of development.*

Harrison refers to 'fragile lands', which he argues is similar to the conception of fragile environments utilised at the United Nations Rio Earth Summit in 1992, and to that of another author, Deneven, writing about environmental fragility in Latin America. The latter emphasises that such fragile lands should be managed according to traditional land-use systems. In essence he argues that these fragile environments impose specific limits to development on their respective communities. However, it is unlikely that traditional land-use systems will enable the inhabitants to develop economically in any systematic fashion. It is highly unlikely that the 'sustainable agriculture' advocated will enable the inhabitants of such areas to join the still rather exclusive club of holidaymakers.

Of course, there can be sound scientifically-based reasons why particular environments may be ill-suited for particular developments. However, as Urry and MacNaughten have pointed out, popular appeals to the environment only occasionally take as their primary point of reference scientific enquiry. Rather, they are more often to do with 'specific contestation about instances of nature which [have come] to symbolise a wider unease with the modern world'. Such sentiments, they argue, reflect an 'aspiration for more meaningful collective engagement and moral renewal'.[9] The moralisation of tourism lies within this trend. The contestation of particular environments, and how they should, or should not be used for tourism, often takes an a priori circumspect view of development. This is presented as cultural sensitivity towards the host community, and hence also as a more moral type of behaviour.

It is also worth noting that there is a lack of coherence in the environmental claims even in their own terms. The moralised view of tourism often holds dear the *specific* environment of the host community, above a broader view of the global environment. More often than not New Moral Tourism involves long-haul travel – this is as true in the green travel guides as it is with the commercial niche market tour operators. Yet other environmentalists, such as Friends of the Earth, highlight air travel itself as the most environmentally damaging aspect of modern tourism.

Fragile cultures?

Moreover, Deneven also argues that, 'social fragility, in terms of organisation, markets, prices, incomes, social relationships and politics ... can be more critical than environmental fragility'.[10] Development is problematic not simply in relation to the environment, but also in relation to the existing balance *between people and the environment.* Similarly, writer on ecotourism Erlet Cater refers to the potential for even the most sensitive of tourism developments to bring problems to 'delicately balanced physical and cultural environments'.[11] This is a crucial point in the environmental critique of tourism development – that there is an existing balance that embodies something positive not just about the environment, but also about culture. Environmental preservation is justified not just for its own sake, but with regard to this relationship. Hence the environmental critique of tourism is at one and the same time a critique of its cultural effects, with culture viewed as this existing relationship between people and nature.

This approach is underwritten by the United Nations, whose Agenda 21 documentation asserts that,

> [indigenous people] have developed over many generations a holistic traditional scientific knowledge of their lands, natural resources and environment. ... In view of the interrelationship between the natural environment and its sustainable development and the cultural, social and physical well-being of indigenous people, national and international efforts to implement environmentally sound and sustainable development should recognise, accommodate, promote and strengthen the role of indigenous people and their communities.[12]

The quote makes explicit that it is the relationship between local (in this case indigenous) communities and the natural environment on which they rely that is central to development. Formulations such as these suggest that nature and history take precedence over technology and aspiration for equality in sustainable development. The application of modern technology, and the levels of development required to make inroads into inequality and poverty, may not rest easily with this deference to indigenous knowledge and local culture, a deference strongly evident in the moralisation of tourism. Local culture – people and their relationship to their immediate environment – is regarded as fragile in the face of development.

'Fragile environments', then, embodies the notion of fragile communities – the people who live within these environments. The relationship between fragile environments and fragile communities is typically discussed as follows:

> Just as traditional uses of soils, waters, plants and animals – often developed over centuries (or longer) of experimentation to minimise change in communities' biophysical life-support systems – may be rapidly

degraded by external influences, the communities' societal structures are equally susceptible to change by external human forces, whose magnitude and potential impacts are not always predictable.[13]

In relation to tourism development, cases where tourism has caused social and economic dislocation are well documented. The problem, however, is that reference to fragile communities suggests that development *as a category* is destructive. Hence the preservation of existing social and economic patterns becomes something intrinsically desirable in the face of this fragility. The discussion of fragility effectively creates a vicious circle of fragility, or perhaps more accurately, poverty. Poorer regions in the Third World, often based upon subsistence agriculture, are inevitably going to be affected greatly by development. Yet in order to overcome their fragility, development – thoroughgoing, transformative development – is precisely what is required.

The anti-developmental content of the language of fragile environments and cultures is evidenced in a discussion of the Annapurna Conservation Area Project in Nepal. Two authors comment that 'village youths are easy prey to the seductiveness of western consumer culture as tourists are laden with expensive trappings such as high-tech hiking gear, flashy clothes, cameras and a variety of electronic gadgetry'.[14] They argue that the Nepalese culture is in grave danger of corruption by tourists. Yet what is really so wrong in aspiring to own a camera and wear fashionable clothes? In fact the real problem here is that for so many Nepalese, like other Third World peoples, the prospect of owning a camera is a distant one because of the poverty that prevails there. The forms of 'sustainable tourism' advocated by the authors are designed to leave these societies as they are – culturally authentic, but grindingly poor.

Further, the authors of this research argue that in the past, 'basic needs have been accommodated by the resource base'. They go on to argue:

> numerous indigenous systems evolved to manage natural resources. Although not perfect, the systems have helped to maintain the quality of the Annapurna environment. Over the last two decades, the explosion in trekking tourism has upset this delicate ecological balance and has contributed significantly to a loss of cultural integrity in the Annapurna region.[15]

Limits imposed upon the indigenous people are not a product of their environment or their culture, but of *the relationship between the two*. In this way, a carrying capacity that precludes transformative development is championed as arising from a holistic approach.

Of course, there will always be situations in which resentment occurs. The point is, though, that this has little to do with tourism development *per se*, but is more likely to be an expression of broader inequalities. Tensions can occur in Nepalese communities arising from, for example, the exposure of

Holiday snaps – *The Beach*

Alex Garland's *The Beach* is a compelling read, and its theme strikes a chord. The beach in question is a place far from civilisation, inhabited by a self-selecting band of western youth. They live out a dream on their own Beach, hidden away from the rest of the world. But the dream turns in on itself.

Especially since the release of the $50 million movie version, starring Leonardo DiCaprio, *The Beach* has become something of a cultural reference point. The story behind the film's making is equally telling. The filmmakers have been accused of damaging the environment on Phi Phi Leh, the small Thai island used for filming. The removal of some existing vegetation (giant milkweed, sea pandanus and spider lily, which then died in a plant nursery before it could be replaced) and the planting of palm trees in its place (removed after filming) has been a focus of the environmental protests that accompanied the filming.

Protesters have brought a legal case against the Thai Forestry Department, the Agriculture Ministry, the Thai film agent and Twentieth Century Fox. The protesters donned DiCaprio masks and chanted slogans to make their point.

Others have argued that Phi Phi Leh was already suffering some ill effects from tourism, and that the film crew cleared rubbish away prior to filming. Evidence of this is to be found in the Rough Guide, which warns that 'unfortunately the disregarded lunch boxes and water bottles of day trippers now threaten the health of the marine life' in the area. Prior to filming, an environmental audit was carried out. After filming, Reef Check, a UN-endorsed coral conservation group, commented that there was no evidence of damage. Yet despite the efforts of the filmmakers, and the stated environmental commitment of the film's star, Leonardo DiCaprio, the film remains tainted by allegations of environmental insensitivity and a $2.6 million lawsuit.

The Beach has become a metaphor for the vain search for spiritualism away from western society. So the argument goes, when too many seek out the beauty of a remote place, it ceases to be remote and loses its beauty – for *The Beach* read Goa five years ago, or even Torremolinos ('a small fishing village ruined by mass tourism') twenty-five years ago. Now environmentalists are worried that the film will popularise travel to Phi Phi Leh and other Thai beauty spots and threaten the natural environment.

Yet others have pointed out that although the palm trees are not authentic the island is cleaner than before. And inhabitants in the region have been far keener to benefit from the commercial spin-offs from the film than to protest.

Nepalese youth to trekkers exhibiting a level of wealth unavailable to them. But to cite tourism as the cause of such problems is to take a very restricted view. Broader inequalities are neglected in a discussion that is essentially about people's behaviour. The defence of cultural difference (from tourists) is prioritised and the aspiration for equality is the casualty.

Perceptual limits?

Carrying capacity can be seen as perceptual rather than based on an environmental imperative. It can be viewed as the numbers of tourists and amount of development possible before *perceptions* of the environment are tainted in the eyes of tourists or residents. There is a commercial issue here – it may well be in the interests of the host to promote a particular type of tourism involving a high spend, which may involve restricting building and tourist numbers. However, there is no evidence that in general hosts prefer fewer tourists, even where tourist numbers are very high. Malta, for example, a small island state and one of the most densely populated countries in the world, attracts large numbers of tourists on package holidays. Yet opinion polls on the island indicate that the Maltese would welcome more tourists – they are seen as a boon economically, and far from animosity, tourists are welcomed.[16] Perhaps this positive view is due to the fact that the island has enjoyed consistent development and increases in income alongside the growing influx of tourists. The Maltese experience suggests that perceptual limits, or at least those perceived by the host societies, cannot be understood outside the broader context of development and prospects for development.

However, the New Moral Tourism often cites perceptual limits as given features of communities under scrutiny. It is certainly true that one can find cases where communities have raised objections to plans for tourism development on the basis of its impact on their livelihoods or way of life. However, this is as likely to be an objection to the partial or limited nature of developments, as it is to too much development or too many people.

One model often cited in the academic discussion of tourism's impacts is Doxey's Irridex. The Irridex provides a linear scale on which to gauge the host's level of acceptance of tourists. It suggest that over time, and with greater numbers, a higher level of tension occurs – that there is a carrying capacity based on the willingness of the host community to accept large numbers of tourists. Initial 'euphoria' with the growth of tourism becomes 'apathy', followed by 'annoyance' and finally 'antagonism'.

In fact what is most striking about the model is not its limitations as a linear model for something as complex and multi-dimensional as a cultural encounter, but that a cultural encounter should be problematised in this way in the first place. The level of tension between different cultures is the starting point of the model. The model suggests that 'a few people are OK, more people are increasingly a problem'. Yet anyone who has visited a popular holiday destination in a bad year when bookings are low will know that the opposite can be true. The Irridex typifies the a priori assumption of the host culture as fragile, and the consequent desire to establish perceptual limits.

The Malthusian outlook of New Moral Tourism

The concept of carrying capacity is applied in a similar fashion in debates on population control, a fashion that could be described as Malthusian. Malthus, writing in the nineteenth century, believed that global population growth would outstrip the growth of the resources needed to provide for that population leading to famine. Today there is often, particularly in the Third World, perceived to be too many people in relation to a limited resource base. Greater population, the Malthusian position has it, puts a strain on the natural environment that will be manifested in its degradation, with in turn severe implications for the survival of populations. However, contrary to Malthus, societies have tended to uncover and harness far greater resources as populations have grown, and, it can be argued, precisely because they have grown, facilitating a greater division of labour within societies. So whilst Malthusian analysis tends to see 'people as problem', as a drain on resources whose growth cannot keep pace with increasing numbers of people, there is much evidence to support seeing 'people as solution'.[17]

The application of carrying capacity to tourism development has a Malthusian character, be it more often applied at a local rather than global scale. Tourism growth is typically viewed as inherently problematic in relation to a fixed stock of resources. Yet the ability to harness and develop resources has been vastly enhanced by increasing amounts of tourists and tourist revenue in many resort areas. More people may pose environmental problems, but they also create possibilities for the resolution of these problems, at greater levels of development and with more opportunities for people as the result.

One small example of this principle is the beach at Benidorm. The much-derided resort of Benidorm caters for tens of thousands of tourists in a relatively small area. Yet it enjoys exceptionally clean beaches. The revenue from tourism enables the local authority to finance keeping the beaches clean, maintaining them as a resource. One could make a similar case with regard to drainage systems and other facilities in many other successful Mass Tourism resort areas, which have often been upgraded to the benefit of hosts and tourists alike through wealth generated by the tourism industry. More people, it turns out, help finance a solution than simply pose an environmental problem.

Closer to home, the dredging of thousands of tons of sand to build up the beach at Bournemouth on England's south coast has attracted many tourists and sustained a growing industry. Increased tourism revenue has helped to finance improvements for tourists and residents alike. Increased numbers of visitors have been instrumental in improving facilities in the town. However, horizons are typically lowered in the Third World, where limited developments, based on 'natural capital', are favoured as more environmentally sensitive in the New Moral Tourism outlook.

In a world in which tunnels under the Channel are possible and artificial islands can be built to accommodate air travel in Japan, one may have expected a more upbeat assessment of society's ability to develop the environment to provide a better life. However, the elevation of nature above development is intrinsic to the designation of rural parts of the developing world as 'fragile'. One author makes this point succinctly:

> Our present admiration for untouched nature, increasingly tinged with guilt about our neglectful stewardship of its dwindling reserves, is a luxury belonging to a fairly advanced stage of social development. Before this had been achieved, nature struck people, if not as enemy, at least as a challenge. Civilisation started by clearing the wilderness.[18]

Yet it is the wilderness that, for some advocates of New Moral Tourism, is seen as fragile, subject to severe capacity limits and in need of *protection* from 'civilisation', which we can in this context read as development.

Conclusion

It is most often the Third World that is presented as under threat from modern development in the form of tourism. It is ironic that so much of the concern over numbers of tourists is directed at their perceived impacts on cultures and environments in the poorest countries, as the Third World

---

Holiday snaps: 'Sustain the Environment and Indigenous Cultures of the Earth'[19] (from The International Ecotourism Society 'Ecotourism Explorer' advice on holiday choice)

'Tourism and travel, as global industries, have many different impacts on both the environment and the cultures of the world. Fragile natural resources like beaches and coral reefs can be destroyed by too many tourists or irresponsible development. Animal habitats can be devastated by visitors, and indigenous cultures can be altered forever by tourists and foreign corporations bringing money and goods.

'No matter where you go or how you travel, you will have an effect on the environment and the people you visit. But travelling responsibly can have minimal negative impact, and, in many cases, can actually help conserve the environment and preserve indigenous cultures.'

is peripheral to international tourism flows. It receives far fewer tourists (and generates fewer still) in relation to land mass and in relation to population size than the developed world. It is, though, the designation of rural environments in the Third World as fragile that provides a rationale for the emphasis on carrying capacity

'Fragility', though, is not given in an environment, but is premised on a limited and limiting view of the potential for development in the Third World, a view that can hold sway when prospects for development have been lowered to the level they are today. The problem is not too much development bumping against natural limits, but too little development to transform societies and push back these limits. Rapacious development from tourism developers and hedonistic tourists with no regard for their hosts, is generally the last of the problems facing Third World countries seeking to better themselves. Rather, it is the partial and limited nature of development that can produce tensions. From this perspective it is at least as true to argue that the problem is not too much development, but too little, and perhaps not too many tourists, but too few.

In the developed world, too, tourism has been described as an industry that 'kills the goose that lays the Golden Egg' – we like to holiday to pretty places, but because so many of us do, these fragile places become spoiled. In the phrase of one report, we are 'loving to death' areas of natural beauty. There is an issue here. Places of natural beauty should be maintained, but for the enjoyment of people rather than as bastions of nature. Mass Tourism has done far more to increase access to diverse environments (natural and man made) than it has spoiled them for others. And elsewhere the Golden Goose is alive and well – last seen strutting Blackpool's Golden Mile.

Walking in the Andes near the border between Chile and Argentina. For some, spectacular natural wonders are a spiritual and even moral counter to urban societies. (Photo: Mick Butcher)

4   Tourists
Too much freedom?

The previous chapter looked at the a priori assumption of environmental and cultural fragility built in to New Moral Tourism. The flip side of this is the assumption that consumers in developed societies have a surfeit of freedom, the exercise of which creates problems in fragile environments. This chapter critically examines the charge that we have too much freedom to travel for leisure, both as individuals and as a society. It suggests that in fact the very qualities that make leisure travel worthwhile are eroded by a circumspect view of individual freedom, and that attempts to morally regulate leisure travel in the name of cultural and environmental sensitivity will only make for guilty tourists and erect new cultural barriers between people.

Freedom under scrutiny

Tourism, especially in its 'traveller' form, has always been associated with freedom. Travel has changed profoundly, but today's backpackers still experience a rare freedom. For many young people, their travels lie between the ages of parental guidance and the strictures of work – perhaps a time for experimentation, radicalism and a little recklessness.

Package holidays are also about freedom. They traditionally carry the image of footloose, carefree relaxation. The holiday is the opposite of work – a chance to leave behind the discipline of working life, and perhaps also the moral strictures of home life.

Yet the innocent notion that you can 'leave your cares behind' seems less straightforward today. Alison Standcliffe from Tourism Concern reminds us that this involves 'closing your eyes to the things you normally care about'.[1] Another author sees the 'increasingly hedonistic philosophy of many people' as militating against making tourism 'sustainable'.[2] Hedonism, once a virtue of tourism, has become a threat. In place of spontaneity, caution and wariness are characteristic of the New Moral Tourism. This cautious approach is captured in a guideline from one of the growing number of codes of conduct for travellers and tourists:

Away from home and free; it is tempting to do things I would never do. I shall avoid this danger by observing myself critically whilst on holiday and behave with restraint. I want to enjoy myself without hurting and offending others.[3]

Freedom is under scrutiny. For the critics of tourism, we have too much of it, both as individuals and as a society. But what is the extent of our freedom to travel? In the more wealthy countries in the world international leisure travel has become commonplace over just a couple of generations. Much of this has been achieved through the package holiday revolution spurred by growing disposable incomes and holiday entitlements. More recently low-cost airlines have also contributed to the ease of travel for footloose young people.

How free are we?

However, tourism's critics present the growth of tourism in apocalyptic terms. Jenny Jones, Chair of the Executive Committee of the Green Party, tells us that a 'staggering' 500 million travel abroad each year. If tourism were a country, we are told, 'it would be the third richest in the world'.[4] In the face of such dramatic pronouncements it is worth reminding ourselves that globally it is the barriers to travel, not its ease, that are most in evidence. It is all but impossible for the vast majority of the world's population to travel freely where they want to, even if they have the income to do so. Immigration controls operate as international pass laws – apartheid in South Africa has ended, but global apartheid has never been more in evidence, even for those wanting to travel simply for leisure.

There is a sick irony in the fact that Third World peoples are often fêted as rich in culture and spiritual depth within their own lands. Tourists can learn much, ecotourists tell us, from more sustainable lifestyles, closer to nature, in the Third World. Third World peoples can be visited and their cultures enjoyed and learnt from – but they can rarely reciprocate. As soon as they want to travel here, they are no longer envied for their spiritual insight, living a simple life close to nature, but are transformed into 'economic migrants' or 'bogus asylum seekers' – a dire threat, we are told, to the western world. Kenyans, when not dressed in traditional costume in the Masai Mara, or protecting wildlife for the enjoyment of well-heeled ecotourists, become part of the mass; a homogenous threat 'out there'. Wives are kept from husbands, children from their parents, by the fortress-like travel restrictions imposed in Europe, the USA and the rest of the developed world.

The expansion of travel has been described as the 'democratisation of travel'. This democratisation has proceeded apace, but still enfranchises only a small minority of the world's population, at least in terms of international leisure travel. Yet for Jost Krippendorf, author of the seminal *The Holiday Makers: Understanding the Impacts of Leisure and Travel*, our freedom to travel 'threatens to engulf us'.[5] If this is true then it paints a dark picture for those who are less free.

Holiday snaps – Happy Christmas, UK-style

In 1991 a plane load of Jamaican tourists coming to the UK were stopped at the airport and turned back. They were over to visit family and friends for Christmas – a commonplace occurrence given Jamaica's expatriate links with the UK. Normally, visas were recommended (although not officially necessary) to ensure a smooth passage through immigration, but on this occasion the British High Commission had told people to travel without, as the visa office was being rebuilt with security screens between prospective tourist and visa staff. The Jamaicans' Christmas was spoiled not by an administrative error, but by a system of checks that treats tourists of a certain colour as prospective criminals.

Modifying our holidays

Tourism's critics are concerned with a kind of freedom. Much of the discussion of tourism today posits natural and cultural limits to its further expansion. Here, freedom is invoked by the critics – *freedom from* tourists is pitted against *freedom to* travel. Tourists are viewed through a dark lens by the critics – always as a threat to the environment or cultures, but rarely as an opportunity.

The critics argue that we should mould our desires, and indeed our lives, around what they consider to be environmental limits. For example, *The Green Travel Guide* argues that the traditional summer vacation is a behaviour pattern borne of habit, bad habit. Rather than spend a summer fortnight of fun and relaxation at a popular holiday destination, we should try to 'break this mould':

> A day or two away from work may prove beneficial, and a longer holiday taken off-season can be much more environmentally sustainable: air traffic is less congested, resorts and other tourist destinations less crowded. Short breaks allow the Green traveller to explore his or her own local environment, without adding to the strains of mass tourism, or to stay at home, whether it is to spend time with our families, catching up with other aspects of our local community lives, or in other forms of recreation.[6]

The implication here seems to be that tourists do not know what is good for themselves, as well as the environment!

The guide goes on to add that increasingly people are 'downshifting' their lives – moving away from a routine nine to five, or even 'supershifting'; living in a way that enables one to choose when one works. One guide to downshifting, written by two journalists who themselves downshifted, includes a section on how holidays fit in to this philosophy.[7] Downshifters can, they argue, turn their diminished income into a virtue by holidaying closer to home or volunteering for environmental working holidays. The package tour operators that one may have booked with when one had less time, pre-downshifting, are predictably criticised as destructive to the environment.

Concepts such as downshifting shed some light on the changed working arrangements for a minority of people, often those able to work more flexibly due to technology or wealthy enough to sacrifice earnings in favour of leisure. But technology has not transformed the working day for most people; often people do the same job at the same desk, but with a computer on it, 'empowering' them to be 'multi-skilled' and 'flexible' enough to work harder, sometimes for longer. And many work hard to afford a few luxuries such as a holiday. Ironically, taking a broader view of labour market changes over the last twenty years, the people who have the most flexibility with their time – the mass of young people channelled into burgeoning education programmes – are as likely to find themselves scraping together the cash for a once in a lifetime inter-rail around Europe or a fortnight in Turkey.

Jenny Jones of the Green Party also advocates holidays at home, which are, of course, more environmentally friendly. Drawing a link between environmental protection at home and abroad she begs the question:

> How many residents of Newbury have been there? If more of them realised what they are about to lose local opposition to the bypass would be insurmountable. Will they instead allow the road to be built, and eventually use it to drive to Europe, to spend a holiday somewhere in the countryside not unlike what is being trashed on their doorstep?[8]

It may be of interest to the residents of Newbury, a town in Berkshire in the UK, amongst others to note that this author speaks from the privileged position of having 'been there – done that'. In the byline at the end of the article she mentions that she has worked and travelled in Jordan, Syria, Israel, Turkey, Crete, Ethiopia, Cyprus, Egypt and Abu Dhabi, as well as having lived in Seychelles and Lesotho. A case, perhaps, of 'do as I say, not as I do'?

Also this critic has missed the point about youth travel. Walk down the main thoroughfare of many cities around the world and you will often see young travellers – only they are all from overseas. And this is the point. Travelling is appealing to many young people not for the cultural experience or heroism, but simply because there is something about being young and *abroad*. With the restrictions of family and school behind you, and the rigours of work ahead, the chance to travel often falls at a transitory period in people's lives. It is hard to feel liberated from these pressures travelling around Berkshire . . . but the idea of being abroad has an entirely different appeal.

Ethical codes: moral regulations for the tourist

The advocates of a New Moral Tourism are not only concerned with the industry, but are profoundly interested in the tourist – your behaviour on holiday. The last five years have seen the emergence of a plethora of codes of conduct to advise and prescribe with regard to personal behaviour. The sentiment behind these is well expressed in the foreword to the recently compiled *Green Travel Guide* written by Robin Pellew of the World Wide Fund For Nature (WWF):

> increased leisure and available money for travel, by train and air, opens up . . . remote parts of the world to the adventurer and imaginative tourist. Faced with these realities, it is necessary that codes of conduct for . . . visitors to foreign countries and different cultures are set in place to ensure the minimum impact on the environment and a maximum sensitivity to the local population and their ways of life.[9]

New opportunities borne of increased travel possibilities have accompanied a growing consciousness of environmental and cultural risks, which these codes seek to address at the level of individual behaviour.

Below is outlined a selection from codes of conduct relating to the behaviour of travellers and tourists.

Tourism Concern

Tourism Concern is probably the British campaigning NGO with the strongest profile. Alongside their numerous campaigns and educational resources, they also issue a general statement of 'Tourism Dos and Don'ts'. Tourists are advised to: 'Save precious natural resources'; 'Support the local trade and craftspeople'; 'Always ask before taking photographs or video recordings of people'; 'Don't give money or sweets to children'; to 'Respect local etiquette – loose lightweight clothes are preferable to revealing shorts, skimpy tops and tight-fitting wear in many countries. Similarly, kissing in public is often culturally inappropriate'; 'Learning something about the history and current affairs of the country helps you understand the attitudes and idiosyncrasies of its people and helps prevent misunderstandings and frustrations.' Overall, the leaflet states that it is 'Promoting awareness of the impact of tourism on people and their environments'. Tourists should 'Be patient, friendly and sensitive. Remember – you are a guest.'[10]

Tourism Concern have recently established their 'Code for Young Travellers'. What is interesting about this and other similar initiatives is the way that a small number of people can put forward a code of ethics that purports to apply to all young people. Even more intriguing is the way that numerous tourism companies want to identify themselves with such a project. The Rough Guide intends to publish the code in its guidebooks.

Now adventurous young carefree travellers will be confronted with a list of dos and don'ts in the front of their guides. If the guidebooks are, as they have been dubbed, backpackers' 'bibles', then the Tourism Concern Code is the equivalent of the Ten Commandments.

Travelers' Code for Traveling Responsibly (Partners in Responsible Tourism)

This American-based advocate of New Moral Tourism advises travellers as follows: 'Interact with residents in a culturally appropriate manner'; 'Reflect daily on your experiences and keep a journal'; 'Support the local economy by using locally owned restaurants and hotels, buying local products made by locals from renewable resources'; 'Get permission before photographing people, homes and other sites of local importance'.

The emphasis on the *local* is typical – benevolence towards the peoples visited is invariably seen as best accruing at a local level, rather than a national one. The codes also often invite us to consider the protocol of travel photography – how many of us have thought it necessary to ask before photographing sites of importance?

The Centre for Environmentally Responsible Tourism (CERT)

CERT was founded in 1994 by a couple on safari who were appalled at a tiger being burnt in a fire caused by a carelessly discarded cigarette. It exists to promote tourist and industry behaviour that helps to protect the environment. One of their three main aims is to 'promote environmental awareness among travellers and holidaymakers'.[11] Whilst the thrust of CERT's campaigning is aimed at the industry, there is also an emphasis on the consumer as an agent for environmental responsibility. Included in a list of advice for the traveller are the following: ' "When in Rome . . .". Respect local customs and sensitivities and follow high standards in courtesy'; ' "Stay on track". Off-road driving, off-track trekking and off-piste skiing can damage sensitive soil and vegetation irreparably. Keeping to marked paths and designated routes minimises the impact of tourism'; '"Don't trade in extinction". Buying products made from endangered species threatens their existence. The sea turtles of Sri Lanka have been brought closer to extinction because of tourists buying turtle products'; ' "When on the beach". Buying shells and hence encouraging the seashell trade, damages marine ecosystems and is nearly always unsustainable. If you are near coral, remember this is made up of millions of tiny animals and takes centuries to grow. Treading on coral or anchoring your boat to it can cause long-term damage.' Tourists are also advised to learn about the local culture as this will 'help you become a more sensitive traveller'.

CERT's educational work is exemplified by a 1997 campaign, in conjunction with Her Majesty's Customs and Excise and British Airways Environmental Branch, asking children to design a poster encouraging tourists to shun souvenirs made from endangered species. Given the abject state of the economy in some Third World tourism host areas, one wonders whether this campaign was more sensitive to wildlife than to people trying to make a living under difficult circumstances.

Friends of Conservation

The Friends of Conservation, an Anglo-American organisation, with offices also in Kenya, issues a 'Conservation Code' for tourists. The preservationist emphasis is clear in the opening lines of the code: 'Tourism is the world's largest industry. It can play an important role in maintaining indigenous cultures and is an invaluable source of foreign currency for many African countries. With a little consideration you can help to preserve this unique part of the world for future generations.'[12] Above all, then, the tourist should be involved in preservation of uniqueness with regard to culture and environment. Whether this is to the benefit of future generations of the inhabitants or future generations of tourists is unclear.

The code proceeds to list a number of dos and don'ts for tourists: 'On Safari'; 'At the Coast'; on 'People and Customs' and 'Shopping'. The advice includes the customary warnings regarding not dropping litter and not wearing clothing that may offend. It also emphasises the illegality of buying souvenirs made from coral, rhino horns, elephant tusks and shells. Friends of Conservation are involved in education not only of the traveller, but also of the host population, and their listed projects include 're-creation of habitat' and 'Anti-poaching support'.

Survival International

Survival International is an organisation devoted to supporting the rights of 'tribal peoples' around the world. Their leaflet advising travellers is titled 'Danger: Tourists' and outlines a number of dos and don'ts. Survival argues with much justification that, 'All too often tour operators treat tribal peoples as exotic objects to be enjoyed as part of the scenery.' Yet Survival's own view is that the integrity of the tribal people's way of life is paramount, and development is therefore regarded as destructive. In this vein they warn that, 'Tourism may distort and irreparably alter the local economy. Tribal peoples who were once self-sufficient or depended on local trade may now become dependent on the tourist dollar.'[13] The preservationist emphasis in Survival's literature not only protects the people from the outside world, but also from any prospect of material development.

American Society of Travel Agents

The influential American Society of Travel Agents (ASTA) has produced 'Ten Commandments for Ecotourism'.[14] The code contains the following advice: 'Respect the frailty of the Earth. Realise that unless all are willing to help in its preservation, unique and beautiful destinations may not be here for future generations to enjoy'; 'To make your travels more meaningful, educate yourself about the geography, customs, manners and cultures of the region you visit. Take time to listen to the people. Encourage local conservation efforts'; 'Learn about and support conservation-oriented programmes and organisations working to preserve the environment'; 'Whenever possible, walk or utilise environmentally sound methods of transportation. Encourage drivers of public vehicles to stop engines when parked.'

The assumption in the ASTA code is that of New Moral Tourism – that the environment – cultural and natural – is delicate, and we should organise our leisure lives around it, preserving diversity. Their Ten Commandments are exemplary of the desire to present conservation holidays as a moral standard rather than an individual choice.

It is notable that although codes of conduct are primarily directed at the tourists, some seek to encourage 'ethical' behaviour by their hosts too. The Annapurna Conservation Area Project (ACAP) *Minimum Impact Code* asks tourists 'to respect Nepali customs in your dress and behaviour' and also to 'encourage young Nepalese to be proud of their culture'.[15] Such advice is commonplace – the tourists are encouraged to see themselves as supporters of the local way of life, even when the locals are deserting it.

Jost Krippendorf also argues the need for education for mutual understanding in his seminal *The Holiday Makers* on the grounds that each group, tourists and hosts, needs to become aware of the other's expectations and needs. Krippendorf puts forward his idea for a more 'human' tourism including proposals to 'inform the host population about tourists and the problems involved in tourism', to 'encourage holidaymakers to try new experiences and behaviour, and to learn how to travel and to prepare', and to 'educate people for travel'.[16] He even advocates that this education should be formalised from the classroom to the travel agent.

The idea that there needs to be a prior written advice for hosts and guests in this way suggests that each party not only speaks a different language, but that they inhabit different worlds. The notion that human conduct requires rules of engagement in the fashion put forward by the codes is striking. Clearly the advocates of codes do not regard hosts and tourists as capable of negotiating their own way through cultural differences and moral dilemmas, which, after all, is surely part of the experience of travelling.

The logic of the angst-ridden critique and the growth of well-meant advice is that tourism should be *for* something . . . something more, that is, than enjoyment and relaxation. In this vein, one advocate of ecotourism demands: 'tourism remains a passive luxury for thousands of travellers. This must change.'[17] In place of luxury, New Moral Tourism offers harsher experiences. Suggestions offered from the Green Party in the UK include learning coppicing, organic gardening, vegan cooking and blacksmithing. Tour

Holiday snaps – excerpts from the codes

'Antarctic visitors must not violate the seals', penguins' or seabirds' personal space' – 'Antarctic Travellers Code'
'Travel like Ghandi – with simple clothes, open eyes and an uncluttered mind' – 'Tips For Responsible Travellers', responsibletourism.com

operators attuned to the New Moral Tourism offer more exotic versions of the same.

It is no longer good enough to travel footloose and fancy free. Tourism is accompanied by constant warnings to limit one's behaviour and to be ethical. From the environmental group Arc's pamphlet 'Sun, Sea, Sand and Saving the World' to Friends of the Earth's advice to question whether you 'need' to travel at all, tourism is now the terrain of moral codes and not a little guilt-tripping. The often impulsive and reckless desire to strike out across Europe or further afield is no longer a good enough reason. A love of music, dancing and drinking provide no defence against the charge of immoral tourism. Excess moral baggage has to be lugged around if tourism is to be acceptable to some. Prominent Green campaigner George Monbiot sees tourism as desirable only when it is carried out by the few who are socially responsible:

> travelling . . . shapes many of those who become the social reformers, the human rights activists without whom every nation on earth would have succumbed to barren dictatorship. These are among the few for whom travel does broaden the mind, for whom exposure to injustice abroad may lift the veil from injustice at home, for whom the conditions suffered by the oppressed of the world, once seen, cannot be tolerated. Whilst they number as tens among the millions, their enlightenment surely means that tourism, for all its monstrosities, cannot be condemned.[18]

For Monbiot, then, tourism is about social reform rather than enjoyment. Those who travel in a carefree and footloose manner are to be condemned for their lack of enlightenment and the damage they cause to the environment and other cultures.

In similar vein Mark Mann, a trenchant critic of Mass Tourism, advocates in his *Community Tourism Guide* that tourists should 'go deeper into local culture, engaging at a deeper level with local people and their needs'.[19] The language used here is that of the social worker rather than the holidaymaker. The need to 'engage with culture', 'respect diversity', 'support communities' and 'make a difference' has replaced fun, relaxation, hedonism and adventure in the vocabulary of the New Moral Tourist.

Codes of conduct can, of course, contain a fair degree of common sense. In this respect, it is less the advice and more the impulse behind formalising common sense that is notable. In advising tourists to respect local culture, there is clearly the assumption that tourists do not do this. In advising tourists to dress appropriately, there is the assumption that such advice is merited for many of us.

The formalisation of common sense through codes of conduct is exemplary of an important trend, that the negotiation of new countries and new cultures is increasingly presented as inherently problematic. The tourist must be on their guard at all times in case they should cause offence. They should interact with local cultures, but at the same time maintain a respectful distance from these same cultures. Even whether or not to give money to beggars has come within the realm of the codes. This problematisation of basic human functions is at times striking – it is easy to forget that we are talking about something as prosaic as holidays.

It is worth noting that the taking up of the codes by eco tour companies, and even by purveyors of Mass Tourism, marks a significant departure. Rather than satisfying customer needs, sections of the industry are increasingly concerned with telling their customers what those needs should be. The sovereignty of the individual has been challenged by the assumption of the ignorant individual, in need of moral enlightenment from a plethora of environmental NGOs and even from tourist guides and tour operators.

In a sense the code-makers treat tourists and travellers like children – unable to think and act as autonomous adults. The assumption of the traveller as childlike threatens to demean the positive function of travel. It shields the subject from confronting and dealing with problems for themselves. It eschews the taking of risks, and hence diminishes the possibilities for learning from our own mistakes. It limits freethinking by contributing to a climate of restraint. Whilst travel takes us to other places and other cultures, our own risk-averse, wary culture accompanies us, lest we should forget about it, let go and start to have fun.

The codes of conduct can also treat the host as a victim – those affected by incoming tourism are presented as victims of a cultural encounter. There is rarely any attempt at a cost-benefit analysis, weighing up the overall pros and cons of development projects. Rather, negatives are given priority above positives, losers over winners, victims over beneficiaries. In addition, counselling local people on how to conserve their own environments contributes to a conception of less-developed countries as in need of paternal care.

Of course the impulse towards codes of conduct is not specific to tourism. The advice industry has boomed in many areas of life such as personal health, parenting and diet. Everyday aspects of life – what we eat, how we bring up children – are increasingly imbued with a pervasive angst and uncertainty. It is no surprise that holidays should fall within this trend, too. Taken together, this 'Culture of Fear' as sociologist Frank Furedi terms it, amounts to a challenge to our moral autonomy.[20] The new emphasis on ethical tourism implies a need for moral regulation at the interface of the tourist and their host. When decisions that may affect others confront us on our travels, whether it is tipping, taking photographs, how to dress or who to purchase from, we have already been told what the answer should be. Hence interpersonal conduct is presented as inherently problematic, and the codes reflect an impulse to police this terrain. Freedom, adventure and hedonism are being eroded by this culture of caution.

As well as the codes, Good Tourist guides have emerged as a counter to Mass Tourism, promoting new forms of tourism such as ecotourism, responsible tourism, nature tourism and green tourism. Three of the most prominent of these are *The Good Tourist*, *The Green Travel Guide* and *The Community Tourism Guide*. They show little restraint in their moral condemnation of Mass Tourism. In *The Good Tourist*, the authors suggest that before going on holiday one should 'Environmentally Audit Yourself'. Questions to be considered in this personal audit include asking yourself: 'Why go on holiday? If you really don't want to go away, don't! . . . Don't follow the crowd.' It urges: 'Get to know the locals . . . it is surprising how, despite the tourist hordes, the desire to be hospitable to guests survives.'[21]

Mark Mann, author of *The Community Tourism Guide*, is quick to differentiate the holidays in his guide from 'the bland facade of mainstream tourism' and the 'tired tourist treadmill'.[22] The guide urges us to 'forget tourism's escapist fantasy and accept that our holidays take place in the real world – and have a real effect on real people'.[23] Escapism and hedonism are replaced by a self-conscious concern and wariness on holiday in the New Moral Tourist's view.

Stark condemnations of tourism, a pursuit that for most people is fun, exciting and carefree, are commonplace. Mass Tourists appear to be neither enlightened about their destructive impact, nor, given the continuing draw of package holidays, willing to engage in a programme of moral rehabilitation offered up by the purveyors of the New Moral Tourism through codes of conduct and a surfeit of well-meant advice.

Whose ethics?

In fact the attempt to provide guidelines for individual conduct on holiday is misguided. The advice offered is often derived from a *particular* ethical outlook, one that stresses the pre-eminence of nature over development, but that is then presented as a universal set of rules for all. Tourists to the Third World may seek to spend their holiday cash in a local community rather than in hotels. This may yield some limited benefits for the rural community, but cut into the service economy in the towns and cities. Refuse to buy a coral necklace and you may contribute to coral preservation, but the vendor may be a little poorer as a result. If we campaign against golf courses on the basis of their use of water supplies, we may conserve the latter but deny people the income from high-spending golf tourists and the consequent possibilities for improving infrastructure. There can therefore be no rules or codes that apply generally. Individuals may decide differently in different circumstances for a variety of reasons – these are matters for consideration and debate. To try to regulate these decisions by drawing up ethical guide-lines is a foolish task. It suggests surrogate parenthood – something young travellers perhaps felt they were leaving at home. Even more foolish is to take conservation as the starting point, when in the Third World societies to which the codes most frequently are applied, *development*, not conservation, is such a pressing need.

No one today is arguing for more freedom for people to travel. There is a sense amongst the critics that we have already surpassed society's limits. The implications of this are that leisure pursuits that many take for granted in the developed world are not to be extended any further. Freedom is viewed in its negative form – freedom *from* tourism is the motif of many of the critics. The result of this is that as individuals we are increasingly subject to a surfeit of well-meaning advice. Some of it is banal and patronising, and elsewhere its claims to be ethical are highly dubious. The traveller seeking adventure is circumscribed. The sun-worshipper is frowned upon. The fun lover is reminded that they are complicit in a 'destructive industry'. Leisure travel has never been subject to such moral proscription.

A Kathakali dancer in Trivandrum, in the Kerala region of India. 'Katha' means story and 'Kala' means performance. The performance, in this case, is for the tourists as dancers gravitate towards the cameras. Tourism has helped to revive the dying art of the telling of ancient myths and epics through Kathakali. (Photo: Mick Butcher)

5 The cultural sensibilities of the New Moral Tourist

New Moral Tourism is preoccupied with 'culture'. Cultural diversity is deemed to have been diminished by modernity and its handmaiden, Mass Tourism. This chapter looks at the way a conception of culture generated in more developed countries colours the way western New Moral Tourists interpret their leisure travel experiences. Put simply, there is a distinct dis-illusionment with modernity in the west that is superimposed upon countries yet to benefit from the modern.

Furthermore, the chapter suggests that ironically the conception of culture implicit in New Moral Tourism, with its emphasis on *otherness*, restricts the very thing the New Moral Tourist holds dear – the ability to learn from and empathise with one's hosts.

The question of culture

Central to the advocacy of New Moral Tourism is the question of culture. A wide-ranging body of literature has emerged in relatively recent years high-lighting the 'cultural impacts' of both tourism developments and the tourists themselves. This is paralleled by the adoption of cultural impacts as the primary argument against Mass Tourism by critical commentators, acade-mics and concerned campaigners alike.

Tourism development is often considered to have ridden roughshod over environmental objections and over the natural landscape itself. The loss of diversity arising from this is of great concern to environmentalists. Develop-ment is also accused of riding roughshod over cultures. Again, allegations that tourism has acted in this way, leading to a loss of diversity in the *cultural landscape*, are prominent.

However, the discussion of cultural impacts is not wholly distinct from that of environmental ones. As we have seen in chapter 3, the *environmental* critique of tourism is often presented through a discourse focussing on *culture* and *community*. In this discussion, culture is presented as embodying the relationship between people and their particular natural surroundings often in a static fashion – change to one or other is eschewed as damaging to both.

The New Moral Tourist places great emphasis on culture. Culture is a dilemma for them though. It is to be respected, supported . . . but not to be encroached upon or patronised. Whilst tour companies and tourists are often accused of imposing their own 'western' cultures and values on to destination communities, New Moral Tourists strive to be culturally equipped and sensitive. But what is meant by culture in the context of a discussion that at times seems like a moral minefield?

The context of the 'culture' discussion is the profound sense of disillusionment with modern society shared by the advocates of New Moral Tourism. In the more developed economies, which are also the major generators of international tourism, there exists a widespread disillusionment with modernity and its cultural associations. Simply, these would include industrial cities, high levels of car ownership, shopping malls, McDonald's and, of course, the much maligned Mass Tourism. Critical of their own culture, New Moral Tourists engage in a search for selfhood, one which locates a spiritual centre in the destination.[1] The New Moral Tourist seeks respite from modernity through a temporary immersion in a culture they perceive to be less sullied by modern society.

The comfort associated with Mass Tourism is self-consciously eschewed in this search for authenticity. A rough-and-ready experience is a virtue for the New Moral Tourist, as it signifies that they are people 'who really want to experience the country and its people'.[2] One travel company, The Imaginative Traveller, inform their prospective clients that, 'In our destinations, you will almost certainly have to contend with such things as relative inefficiency, a more relaxed attitude to time, cancellations and closures without explanation, outdated facilities, suspect plumbing and apparently mindless bureaucracy.'[3]

The adjective 'real' is often prefixed to 'culture' or 'people' in the ecotourism and trekking brochures. 'Real' typically turns out to mean 'rural' – what is real or authentic for the New Moral Tourist is not to be found in cities or towns, which remind him of home, but in rural, 'sustainable' lifestyles. Alternatively, Urry argues, referring to a broadsheet newspaper *Campaign for Real Holidays*, 'real' refers to the Romantic tourist gaze, well away from the 'masses'.[4]

A short travel piece titled 'My Journey with the Bedouin' in the *Independent* travel supplement illustrates the rejection of one's own modern society implicit in travel:

> We didn't know if visiting the Bedouin nomads of the Syrian desert would be remotely possible, let alone pleasant. There were worrying stories about overcrowded tourist sites in Jordan, and even worse rumours that many of the trips were a set-up, with the supposed nomads returning to their homes and televisions after the tourists had gone home fooled and fleeced.' [Then followed a cold night disturbed by a snoring camel] 'By now it was getting cold, time to take shelter ourselves and relax and await dinner. As the sun sank below the horizon

I asked myself if there could be any more desirable way to escape the modern world.[5]

In *The Spiritual Tourist*, a travel book written by journalist Mick Brown, the author regards this search for spirituality through travel as 'a symptom of collective uncertainty'. He argues that this malaise in western selfhood finds its expression in the outpouring of emotion over the death of Princess Diana, and in travel as a 'spiritual search'.[6] The author's search takes him from Euston Road to Tibet . . . and back to London. His search yields little that is profound – the search for spirituality proves illusory.

Sociologists have even likened tourism to pilgrimage, or a search for the sacred.[7] Of course, for most tourists leisure is the aim and thoughts of deeper truths are far from their minds. However, the comparison perhaps warrants inspection with regard to New Moral Tourists, whom one might regard as seeking some higher understanding on their travels. But whilst traditional religions have sacred beliefs, customs and places – they are sacred systems, New Moral Tourism, in so far as it has any religious parallels, is definitely New Age. The spirituality sought is something personal and specific to the individual, not part of any system and just as New Age spirituality is not a single moral framework, neither is New Moral Tourism. Rather, it is a fluid phenomenon. It seems that everyone has their own moral world, so the notion that there could be a single 'nature', 'culture' or indeed a single ethics is wrong. Rather, the moralisation of tourism is a tendency to see tourism in moral terms connected to a heightened perception of environmental and cultural risks and a circumspect conception of progress embodied in the growth of mass travel.

In *Bobos in Paradise*, David Brooks observes the Bobos (Bourgeois Bohemians), whom he considers the new elite of the information age and who identify themselves with the values of the counter-culture and with liberal concern. Bobos, at the cutting edge of the knowledge society, are keen to leave progress behind when it comes to leisure pursuits:

> The Bobo, as always, is looking for stillness, for a place where people set down roots and repeat the simple rituals. In other words, Bobo travellers are generally looking to get away from their affluent, ascending selves into a spiritually superior world, a world that hasn't been influenced too much by the global meritocracy. Bobos tend to relish People Who Really Know How To Live – people who make folk crafts, tell folk tales, do folk dances, listen to folk music – the whole indigenous people/noble savage/tranquil craftsman repertoire. . . . Lives therefore seem connected to ancient patterns and age-old wisdom. Next to us, these natives seem serene. They are poor people whose lives seem richer than our own.[8]

Bobo culture extends beyond the knowledge economy elite – the rhetorical rejection of one's material wealth in favour of spirituality is a theme throughout contemporary western culture.

But are the signs of an unmodern existence sought after by tourists important? Do they mean anything beyond romantic fancy? The notion of the
'post-tourist', or post-modern tourist, may suggest otherwise. Urry, Feifer
and others cite this post-tourist, who is associated with playfulness.[9] They
are aware that experiences may be staged, and are not necessarily seeking
an elusive 'authenticity'. For Urry, tourists are 'unsung armies of semioti-
cians . . . Fanning out in search of signs of Frenchness, typical Italian
behaviour, exemplary oriental scenes, typical American thruways, traditional
English pubs'.[10] When on holiday, why not?

Post-tourists, or post-modern tourists, however, may also be post-modern
in the sense of being *against* the modern.[11] Hence playfulness does not
necessarily preclude a moral sense of one's leisure activities – holidays can
be critical statements about modern society and indeed modern tourism.
David Brooks' Bobos on holiday are modern-day *flâneurs*, strutting out and
displaying not just a sense of taste, but also a sense of morality for all to
gaze upon.[12] (Numerous dinner party conversations with 'post-modern'
tourists have convinced me on this point!)

For Brooks' *Bobos in Paradise*, as for New Moral Tourists, not only is
otherness sought after, but there is a sense in which it can be elevated above
one's own culture – as Urry has pointed out, even the mundane can be
considered extraordinary whilst on holiday.[13] The tourist gaze fixes upon
sites that appear to offer an unmodern existence; an existence from which
the tourist feels they have much to learn. The New Moral Tourist gaze is
a gaze in awe.

Many writers prominent in the academic debates around tourism have
echoed this sense of disillusion with modern society. For Dean MacCannell,
author of two classic books on the sociology of tourism, *The Tourist: a New
Theory of the Leisure Class*, and *Empty Meeting Grounds*, we are living in the
'most depersonalised epoch in history', an era in which human relations are
denigrated by market relations.[14] Sociologist Graham Dann speaks of a 'situation of perceived normlessness in the origin [tourism-generating]
country'.[15] Jost Krippendorf, author of *The Holiday Makers*, articulates the
sense of malaise thus:

> they (tourists) no longer feel happy where they are – where they work
> and were they live. They feel the monotony of the daily routine, the
> cold rationality of factories, offices, apartment blocks and transport,
> shrinking human contact . . . the loss of nature and naturalness.[16]

Respite from the modern malaise, it is argued, is unlikely to be found in
mass package travel of the sun, sea and sand variety. Indeed, as an exemplar of modern society, Mass Tourism is roundly condemned by advocates
of the New Moral Tourism. As we have seen, New Moral Tourism is not
merely a question of personal taste, but carries a strong moral condemnation of the activities of the masses on their holidays.

The prevalence of this critical view of modern society, and modern Mass Tourism, is an important contextual feature of today's advocacy of New Moral Tourism. Such tourism can become part of a search for selfhood and identity in a world lacking agreed norms and racked by doubts. The advocacy of 'culturally sensitive' tourism is shaped by this, as is the understanding of culture itself. 'Culture' in this usage is often refracted through a distinctly western lens; one that both elevates the host's culture and at the same time restricts its development. There are three facets to this: first, the host's culture is celebrated as holding things together and maintaining the status quo in a society. Change becomes defined as an attack on culture. Second, culture is rooted in the past, in tradition, rather than being connected to the making of a future. Third, and most vitally for this discussion, culture is seen as what makes people different from one another – culture is read as *cultures*.

Culture as function

Culture is often seen as analogous to the human body, with different organs (cultural norms, perhaps) enabling the overall functioning and survival of the body (the society).[17] Change to aspects of culture wrought by tourism is seen as upsetting the functioning of the society more generally. Culture, then, can be viewed as being functional.

This view of culture is in many ways uncontentious. However, if we are studying societies in the context of social change, such a conception may carry profoundly conservative assumptions. To stretch the analogy, the human body may function biologically, but human beings, and their societies, function *socially*. Likening human culture to biology inclines towards a naturalisation of culture; and that which is natural is not disposed to anything other than the most gradual, evolutionary change.

Of course cultural norms, ceremonies and forms of economic activity do play a role in cohering societies as they are constituted. Yet to see culture as functional in relation to societies *as they are* is to ignore the creativity of individuals, to downplay change. Culture, in this view, can appear to exist outside of, and prior to, the individual – culture makes man, rather than the other way round. Such a view typically attributes different levels of economic development to cultural factors, and these cultural factors in turn persist in an unchanging society. Persistent inequality, wholly undesirable from a humanist standpoint, becomes reinterpreted as 'cultural diversity' and celebrated under the ethical tourist gaze.

In academic discourse, the models sometimes used to conceptualise the relationship between tourists and hosts can be limited in this fashion. For one author, the level of impact arising from the meeting of host and tourist results from, 'the inherent strength and ability of the host culture to withstand, and absorb, the change generators whilst retaining its own integrity'.[18] Change, here, is an exogenous variable to be withstood, and culture has its own 'integrity' viewed in much of the literature as virtuous in and of itself. Following from this, it is no surprise that change is often viewed with a conservative and cautious eye.

The New Moral Tourist sees tourism in this fashion. The host society is often subject to a romantic gaze by the tourist in search of a sense of authenticity – its culture is seen as unsullied by consumerism and embodying spiritual values and a sense of community sought after in more wealthy tourism-generating regions. 'Culture' coheres nirvana, protecting it from the modern assault. On the other hand, cultural transformation is often interpreted as destructive by the New Moral Tourist. They seek timelessness, not change, in a rapidly changing world. In this way a sense of 'culture' is constructed – one that may not correspond to the desire or potential for change in less economically developed regions.

Prominent examples of 'culture as function' are to be found in the discussions of emergent destinations such as Goa in India. Customary references to the 'destruction of' or 'damage to' culture often interpret change as being inevitably in a negative direction. A notable example of this was the BBC's *Our Man In Goa* television programme, presented by Clive Anderson, in which Mass Tourism was presented as the destroyer of Goa's cultural integrity. For Clive Anderson, '[tourists should] find somewhere else to go, with a culture that is not so fragile and with very little of value that can actually be damaged . . . somewhere like Euro Disney'.[19] American culture, implanted into France, has 'very little of value', whilst Goan culture is a valuable, fragile thing, maintaining the integrity of Goa. Clive Anderson's comments are frequently echoed in the pages of travel supplements and New Moral Tourism brochures.

Yet while many commentators may regret 'cultural levelling' here, there is evidence to suggest that many residents of Goa are less precious about culture than some of their western advocates. Journalist Sam Woolaston, in search of Goan culture, discovered Lamani girls with East End accents (mimicked from Bianca, a character on the British soap opera *Eastenders*), and music students at the Panjim academy whose favourite music was Guns and Roses (one of the students was nicknamed Elvis). Evidence of Goan culture was to be found in 'Goan House', a mixture of western House music and Goan instrumentation. Such is the vibrancy of Goan culture![20]

The lifestyles of Goa's inhabitants have changed markedly, but it would be quite wrong to attach adjectives implying destruction to this process, as some observers have noted.[21] Local businesses have developed and adapted to meet the needs of tourists, with benefits, be they limited, accruing to local people. The inhabitants of some 'unspoilt' areas further along the coast have expressed envy at the development taking place in Goa, and a desire for the sort of development that could enable their children to travel, just like the tourists.[22] These hosts aspired to equality at the expense of 'culture'.

Such sentiments contradict the commonplace view that 'fragile cultures' should be sustained, and even 'preserved', but should come as no surprise.

After all, culture in the more economically developed regions has constantly changed as development has transformed the way of life. Notably, travel itself has become part of the culture of the developed world, setting it apart from the less-developed regions – the regions most often seen as being 'unspoilt' and worthy of 'preservation'.

Inevitably 'culture as function' is seen as being disrupted by the growth of commerce connected to the tourism industry. This theme is developed in *Rethinking Tourism and Ecotravel* by Deborah McClaren. The book goes further in advising travellers how to act on their travels to restrict what she sees as harmful commercial development leading to 'the Paving of Paradise'. The author recounts a visit to Montego Bay, Jamaica:

> I tried to meet some local people without being accosted by entrepreneurs, but I was taken to other all-inclusive resorts around the island . . . I noticed the creation of a fantasy tourism culture that by no means represented the real culture of Jamaica.[23]

For this author, business culture is inauthentic – people trying to make money is alien to the 'culture' of the island. But surely the struggle to survive poverty is a central feature of Jamaican culture. In fact, 'being accosted by entrepreneurs' is a constant feature of tourism in less-developed and middle-income regions. It is as much a part of the culture as the 'family scheme' developed by the Jamaican government to encourage 'real' cultural exchange, in which tourists experience traditional family hospitality with no formal payment demanded (although a sizable tip would rightly be expected). The problem is not pushy entrepreneurs, but the fact that this is, tragically, the way many people have to make a living. It is the failure of the formal economy that has necessitated the rise of the informal, 'pushy' economy. Yet thoroughgoing economic development is ruled out of order by the New Moral Tourism lobby, who see it as an affront to the sanctity of indigenous culture.

Culture as the past

The past is a foreign country – they do things differently there.[24]

The modern understanding of the term 'culture' carries a strong association with the past, with history. To be more precise, 'culture' is often used in a sense analogous with a narrative of the past, described by historian E.H. Carr as 'History with a capital H'.[25]

Heritage, the age old and tradition are fascinating to the tourist seeking respite from the modern world. Indeed the pull of the past on the modern tourist psyche has probably never been stronger, leading to the growth of the Heritage Industry seen by Robert Hewison as a sign of decline in the British context.[26] Put simply, the association of the present with a lack of authenticity leads to a search for authenticity in the past. Elements of the past are located in other, lesser-developed, regions and these regions then take on the role of arenas for the personal self-discovery of the tourist. The host culture is held to possess something that the tourist's culture has lost; a sense of community, of spirituality, of being closer to nature. Of course the desire to preserve this past as a 'destination' is rarely combined with a desire on the tourist's part to actually live in it.

One example of the reverence of the past and its moral elevation *vis-à-vis* modern Mass Tourism is the Proyecto Ambiental in Tenerife, a European Union-funded sustainable tourism project. Its promotional literature pledges support to '*traditional communities* under threat', and the guilty party is held to be Mass Tourism development which has, 'devastated rural communities and forced ... an *age-old culture* to the edge of extinction' (my italics).[27] Tourists are invited to assist in sustaining traditional farming techniques and researching the mythologies of the goatherders, whilst no doubt casting a condemnatory (but perhaps occasionally envious) eye down to the brimming resorts beneath them on the coast, and beneath them morally.

For all the consequent problems of Mass Tourism development in Tenerife, there is no doubt that it has contributed to change for the better. The reverence of tradition evident in such schemes as the Projecto Ambiental colours the way the tourist sees the host society. 'Culture as the past' denies the host their creativity. As Hewison puts it, 'hypnotised by images of the past we risk losing all capacity for creative change'.[28] The problem with the moral elevation of the past is that 'traditional communities' become a living museum piece, valued for their authenticity in the way one may value a piece of antique furniture.

Of course, history need not be read as 'History with a capital H'. History can embody human self-development. It can: 'lose(s) its exclusive association with the past and become(s) connected not only to the present *but also to the future*' (my italics).[29] As Hewison and others have argued, though, contemporary society exhibits uncertainty in the present, and little optimism about the future. The past has become a source of comfort in an uncertain world.[30]

New Moral Tourists may seek out the past, but when they engage with the past it is not as history makers – people who, to paraphrase Marx, want to understand the world in order to change it – but 'history takers', building up a stock of inwardly enlightening experiences. Rather than study history in order to understand the world, a claim of the Grand Tourists of the eighteenth and nineteenth centuries, New Moral Tourists gaze on history to reinforce a sense that the world is too diverse to understand at all. They feel humble and insignificant confronted with natural and cultural difference.

An emphasis on 'culture as the past' obscures culture as the making of a future. The history tourists may see as so vital in contributing to diversity may be less appealing for the hosts in poor countries. Traditional ways of life are undoubtedly fascinating to the tourist, who themselves enjoy a way of life that probably has little to do with tradition and owes much to the benefits of a modern existence. For New Moral Tourists a fascination with the past may be a product of disillusionment with the present. For those who live in less-developed regions, the past may be a constraint from which they wish to escape. In this way categories borne of post-modern angst may be imposed upon those yet to benefit from the modern.

Culture as cultures

The term 'culture' has been described as 'one of the two or three most complicated words in the English language'.[31] Historically, two strands of 'culture' are discernible: universal or human culture, and culture as *cultures*. For the former, rooted in the Enlightenment conception of human progress, culture embodies a common human project of social development. For the latter, originating in the Romantic reaction to Enlightenment rationalism, culture expresses human difference. Raymond Williams identifies a distinctly anthropological view of culture, one that falls into this category, thus: 'the specific and variable cultures of different nations and periods, and also the specific and variable cultures of social and economic groups within a nation'.[32] This view posits culture as difference, as *cultures*. It is the cultural outlook of New Moral Tourism.

The elevation of cultural difference above commonality is a key feature of much of the advocacy of New Moral Tourism. Yet as culture is such an amorphous concept, it may be more useful to consider a *sense* of culture. One recurrent theme in the literature is the importance of *difference* to the tourist. For one author, commenting on modern tourism: '(modern man) is interested in things, sights, customs and cultures different from his own, precisely because they are different. Gradually a new value has evolved: the appreciation of the experience of strangeness and novelty . . . valued for their own sake.'[33]

Another author, in an article titled 'Marketing Authenticity in Third World Countries', neatly summarises the cultural outlook of the New Moral Tourist:

> it seems that tourists and the indigenous peoples are incommensurably different within the touristic process, and indigenous people can only continue to be attractive to tourists so long as they remain undeveloped and perhaps, in some respects, primitive.[34]

David Lodge's novel *Paradise News* is insightful with regard to the culture debate.[35] The book follows the fortunes of tourists to Hawaii. Some are sun-seekers, some visiting relatives, and one prominent character, Rupert Sheldrake, is on an anthropological holiday. He spends his time studying the behaviour of the other tourists and warning them of the futility and destructive nature of their leisure.

Sheldrake comments that, 'I'm doing to tourism what Marx did to capitalism, what Freud did to family life. Deconstructing it.' Sheldrake's theory is that the sheer repetition of the word 'paradise' in brochures, in hotels and in the resorts brainwashes the tourists into thinking they really are in paradise.

Sheldrake travels alone – his fiancée ended their engagement: 'She said I spoiled her holidays, analysing them all the time.'

Sheldrake is a fictional character, but definitely a man of his time. He voices grave concerns with tourism's ill effects, which chime with those of many of tourism's critics today. His unease at what tourists do and how they behave is echoed in the steady stream of codes of conduct and ethical advice given out to tourists today. Also, there are more than a few real-life Sheldrakes who divide their time between studying the 'host–guest' relationship in exotic locations, and writing up the results of their observations for learned journals and concerned environment editors. Indeed, some may even be employed by the United Nations and various corporations to examine the cultural impact of proposed developments.

New Moral Tourism is a little like amateur anthropology. Just as the anthropological study of tourism emerged with concerns over cultural contact between hosts and guests,[36] New Moral Tourism has reflected growing misgivings with Mass Tourism.[37] Both anthropologists and New Moral Tourists are interested in learning about the culture of the host. Both may also seek to minimise their own impact on the host's society – anthropologists seek to blend in order to avoid eliciting behaviour different from the norm, and New Moral Tourists may be wary of their own capacity to damage the local culture. Also, neither is satisfied with staged aspects of the host's culture, in which traditional festivals and rituals are presented as spectacles for tourists.[38] Both seek to go beyond that, potentially into the authentic 'backstage' world of their host.[39]

Many New Moral Tourism companies and development initiatives utilising nature-based tourism appeal to this desire to go 'backstage'. Tanzania's NGO-funded cultural tourism programme offers tourism 'the People to People way'.[40] The literature says that the tours 'offer visitors insights into the life – traditional and modern – of Tanzanians at home and at work, at play and at rest'. The brochure is filled with pictures of cultural life – predominantly villagers working on the land and taking produce to be sold, as well as one of a mother feeding her baby. The photographs would fit well alongside an anthropological account of village life in poor, rural Tanzania.

The anthropological conception of culture has a defining characteristic – cultural *difference* is assumed as a starting point whilst *common* aspirations and desires between host and tourist are rarely examined. Arising from this there is an emphasis on, and guarded approach to, cultural contact. There is, in Sheldrake fashion, a tendency to treat the host–tourist relationship as a constant cultural dilemma.

Viewing the host and tourist through the prism of anthropology is characteristic of the critique of tourism. It presents host and tourist as inhabiting two separate worlds, with a cultural divide in between. The anthropological perspective is best expressed by its exponents, the tourism anthropologists. According to anthropologists of tourism Dennison Nash and Valene Smith, 'anthropologists draw on a transcultural perspective that embraces all of the cultures of mankind'.[41] They 'specialise in the study of the dynamics of human cultures and cross cultural communication'.[42] Both of these quotations refer to the study of *cultures* rather than culture. The starting point of the tourism anthropologist is, then, *difference*, as is that of much literature in the field of tourism impacts generally. The concept of acculturation, for example, has as its starting point the formal counter-position of two separate cultures. One culture encounters another, and there is an interaction between the two. This is codified in the definition of acculturation given by the Social Science Research Council: 'culture change that is initiated by the conjunction of *two or more cultural systems*' (my italics).[43] Clearly, if we begin with different 'cultural systems' then common strands of culture may be overlooked.

In the field of tourism anthropology, there are rarely allusions to a common human culture. A common position is that of Spanish academic Nunez, who describes the consequences of acculturation as being 'when two cultures come into contact of any duration, each becomes somewhat like the other through a process of borrowing'.[44] The result of acculturation is a process of mediation between the two cultures. This can result in resistance to the dominant culture by the more fragile, domination of the latter by the former, or a process of hybridisation may occur, through which cultures borrow from each other. Such hybridisation can be seen as a creative process,[45] but the relationship is often seen as fraught with problems, even encapsulating 'cultural imperialism'.[46]

For another author, the level of impact arising from the meeting of host and tourist results from the 'interaction between the nature of the change agent and the inherent strength and ability of the host culture to withstand, and absorb, the change generators whilst retaining its own integrity'.[47] The tourists are the change agent, an exogenous variable to be withstood, and culture has its own 'integrity', viewed in much of the literature as virtuous in and of itself.

With regard to the formal counter-position of different cultures in the above and many other formulations, it is worth thinking about how far one could take such an approach. Does acculturation take place when British tourists go to France, or when residents of the city travel out to the country? In fact, the problematisation of the meeting of cultures usually comes to the fore in relation to 'culture contact and culture change, particularly where the contact has involved a more powerful, more developed Europe and North America and the less powerful, less developed world of countries such as those in most of Africa or Latin America'.[48] The principal cultural divide is typically seen as being, then, between the developed and developing worlds.

Of course, tourism is not always seen as destructive in relation to the host culture. It is sometimes seen as a positive factor when it *reinforces* a cultural practice. On the Greek island of Crete, tourism is seen as holding out the hope of sustaining traditional textile production, if cultural tourism can attract tourists interested in buying such traditional goods.[49] On a similar theme, referring to the impact of tourism on the Masai peoples in Kenya and Tanzania, Dean MacCannell argues that, through tourism,

> the assimilation of primitive elements into the modern world would allow primitives to adapt and coexist and earn a living just by 'being themselves', permitting them to avoid the kind of work in factories and as agricultural labourers that changes their lives forever.[50]

For MacCannell, Masai culture can be sustained, not destroyed, by a degree of commercialisation arising from tourism. It is worthy of note that there is an underlying 'anti-change' assumption here. For MacCannell those Masai who have moved towards work in factories have ceased to 'be themselves' and are the worse for it.

It is notable that arguments both for and against tourism development often emphasise maintaining cultural difference as a goal. The integrity of culture is deemed worthy as an end in itself.[51] Where cultural change is discussed, it is as a result of acculturation, through the formal interaction of different cultures. Such an approach undoubtedly has its merits in helping to identify potential tensions arising from tourism, and much research has been carried out highlighting these. Undoubtedly even small numbers of tourists can have a considerable impact in an area where infrastructure is poor, and where the tourists stand out as being conspicuously wealthy compared to their hosts. However, the assumption of the *primacy* of cultural difference has become an unhelpful dogma within much of the advocacy of ethical tourism. The formal counter-position of cultures, host and tourist, negates approaching the issue from the perspective of *commonality*. That the host and tourist may share common needs, desires and aspirations with regard to development is less commonly considered. Not least of these may be the aspiration to join the growing, but very limited, ranks of the world's tourists, an aspiration only achievable in the context of economic growth and cultural change.

Of course, as differences exist between societies, an appreciation and interest in such differences in cultural values can be a positive aspect of tourism. But somewhere in the rejection of culture as a universal concept the aspiration for *equality* has been lost. For Sir Crispin Tickell, referring to tourism in less economically developed regions, humanity should, 'glory in our differences rather than subordinate ourselves to some grey middle standard'.[52]

But in so far as these differences reflect different levels of economic development, such assertions beg a commitment to material equality.

The conception of culture as cultures has a long tradition in anthropology. The work of the founding father of modern anthropology, Franz Boas, stood in opposition to the universalistic Enlightenment conception of human culture.[53] The Enlightenment-influenced writers understood culture as a universal rather than a particular concept. J.S. Mill, for example, wrote of different cultures, but in the context of a 'philosophy of human culture' and of 'the culture of the human being'. He was referring to the notion that all people, and all societies, are capable of progress, that human development is a common cause and that certain common standards apply.[54] This view held that there was such a thing as societal progress – societies could become more 'civilised'. Many have criticised such a view as Eurocentric. However, the Enlightenment conception of culture upheld the possibility and desirability of political and material equality. In relativising culture by presenting it as *cultures*, progress towards *equality* is contradicted by an acceptance and defence of the primacy of human *difference*.

Morally, this starting point is often justified through a sense of injustice at past and present cultural domination, or 'cultural imperialism' within which defence of difference becomes seen as a counter to domination. Anthropologists, and New Moral Tourists, tend to take the side of the Other. Such an approach is in many senses admirable. Yet it is important to view cultural difference within the context of broader human culture. Otherwise, in countering *cultural* domination, the *material* inequality that characterises different societies and shapes 'culture' ironically can be reinforced through a rejection of development as culturally inappropriate. If, as one anthropologist critical of modern tourism has it, wants are 'culturally derived . . . [and can] vary greatly from one society to another' [55] why should we see material inequality as anything other than a product of cultural difference? New Moral Tourism defends to the hilt the freedom to be different but not, apparently, the freedom to share the same standards, the freedom to aspire towards equality.

French anthropologist Claude Lévi-Strauss developed a perspective on culture that is key to understanding the outlook of the New Moral Tourist (he has also written directly about tourism as a danger to cultures). Lévi-Strauss's anthropology shares many of the themes of Boas's work. For Lévi-Strauss, human thought-processes comprise contrasting pairs of symbolic dualisms – a conception of 'Self' and 'Other' impose upon us modes of thinking from which we cannot escape. For Lévi-Strauss: 'We must accept the fact that each society has made a certain choice, within the range of existing human possibilities and that the various choices cannot be compared to each other.'[56] The 'other culture' remains, then, incomprehensible. Moreover, it is 'impossible' to deduce any 'moral or philosophical criterion by which to decide the respective values of the choices which have led each civilisation to prefer certain ways of life and thought while rejecting others'.[57]

There is an important nuance in the view of the 'Other' in Lévi-Strauss's work. Lévi-Strauss held that 'primitive' cultures were, in fact, a stripped-down version of modern ones, without the trappings of modern consumerism. Hence in the context of a widespread disillusionment with modernity, articulated by Lévi-Strauss (and evident in many contemporary writings on tourism), the developed world has much to learn from its less-developed counterpart. Or, if as MacCannell asserts, 'advanced capitalism is accomplishing the destruction of nature and the human spirit',[58] then perhaps a more basic sense of 'who we are', akin to the tourist trying to 'discover the real me', can be gleaned from societies less advanced and more free from the perceived fetters of modernity.

Lévi-Strauss does suggest a way out of the dilemma – a solution that chimes with many of the micro-solutions evident in the New Moral Tourism discourse. He argues that the ideal situation would be one where

> communication had become adequate for mutual stimulation by remote partners, yet was not so frequent or so rapid as to endanger the indispensable obstacles between individual and groups or to reduce them to the point where overly facile exchanges might equalise and nullify their diversity.[59]

Lévi-Strauss calls on international institutions to see the 'necessity of preserving the diversity of cultures in a world threatened by monotony and uniformity'.[60] This is a rallying cry that institutions and NGOs have certainly heeded since the 1970s, when Lévi-Strauss wrote this. A host of NGOs, as well as national and supranational initiatives, have firmly established cultural preservation as a worthy goal, and codes of conduct and carrying capacities have become a part of a new etiquette around the discussion of the relationship between people who happen to be on holiday and people who happen to live in areas visited by holidaymakers.

The understanding of culture, or cultures, in the writings of Lévi-Strauss mirrors the contemporary debate over tourism. Civilisation, or a common standard of culture, 'is the destroyer of those old particularisms, which had the honour of creating the aesthetic and spiritual values that make life worthwhile'.[61] Lévi-Strauss's sentiment may be reassuring to New Moral Tourists, who celebrate cultural diversity – a diversity they may freely claim not to comprehend. But for societies struggling against poverty there may be little time for aesthetic contemplation. The elevation of the primitive in Lévi-Strauss's writing and in the worldview of the New Moral Tourist can easily cross over into an aestheticisation of drudgery.

In similar vein to Lévi-Strauss, many of today's critics of tourism subscribe to the view that it can fictionalise situations involving cultural contact in a way that, as MacCannell argues, 'assume[s] the superiority of the west'.[62] New Moral Tourism is certainly a response to this perception of cultural arrogance. However, the New Moral Tourist creates their own fiction based on the *inferiority* of the west – they elevate the natural and cultural characteristics of poor host societies and decry their own. Whilst presented as sensitivity to these communities, it is this approach that implicitly restricts the meaningful contact that MacCannell also argues for, and that New Moral Tourists seek, by infusing awe, caution and not a little guilt into cultural encounters between individuals.

A notable point about the assumption of the primacy of cultural differences is the way in which the critics of modern tourism (which is assumed culturally insensitive) present their case as a radical defence of exploited host communities. Yet it is worth commenting that the 'cultures' perspective originated in conservative thought in the nineteenth century as a reaction to Enlightenment humanism. Most notably, Herder's concept of culture, which upheld the primacy of human differences rooted in land and blood, prefigured the racialist doctrines of Gobineau and Knox. Herder's philosophy denied culture as the sum of human material and intellectual progress, in favour of a view of humanity defined by its differences.

Cultural relativism was bound up with the promulgation of the idea of race in the nineteenth century. Today it is invoked as a counter to the perceived globalising mission of western culture. Whilst no one would argue a direct comparison between contemporary cultural relativism and that of a century and a half ago, it is evident that in both cases it seems to negate the possibility of material equality, and directs us away from examining culture from the perspective of commonality. It is not, therefore, an agenda that should be readily granted the moral high ground it has occupied thus far in many quarters in debates on modern tourism.

Community versus development

The presentation of the relationship between host and tourist as a cultural minefield has implications beyond the critique of tourism – it makes assumptions about the desirability or otherwise of development. This can be illustrated by contrasting the commonplace perception of large Mass Tourism developments as being destructive of the host culture with notions of small-scale, 'eco' or 'alternative' tourism. For example, one writer compares favourably the way Sherpas in Nepal have maintained their reciprocal community-based social arrangements in the light of tourism development, compared to the 'impersonal' development in Mijas on the Costa del Sol.[63] The development of the tourism economy on the Costas is seen as negative from a human perspective, perceived as having destroyed an old cohesive community. This is a sentiment shared by some New Moral Tourists, for whom Spanish coastal developments represent the excesses of Mass Tourism and Mass Tourists. The Nepalese Sherpas, on the other hand, are seen as having avoided 'cultural levelling' and maintained their sense of community.

Undoubtedly there is a kernel of truth in this argument. However, the citizens of Mijas are much better off than the Nepalese Sherpas. Many are able to engage in leisure travel, so in this sense can engage in a relationship with the tourist based to a greater extent on equality than the Sherpas can. Indeed, late on in the holiday season the Costa del Sol is popular with the Spanish themselves. For many younger Spaniards, the development of tourism has opened up new opportunities to meet with and relate to a broader section of society than previous generations.

Some may bemoan the changes to community, but perhaps we should see this as a price worth paying.[64] Community can be seen as an enabling concept – a level of social organisation that enables attitudes of reciprocity and tolerance to develop, as people are linked by their closeness and commonality of goals. Yet community can also be divisive. It can be an expression of 'small town' lack of ambition, a bind on younger generations who aspire to a broader sense of themselves in the world. The latter impulse is why travel itself has been so appealing throughout history – it expresses one's ability to go further, experience more, to learn more and to overcome limits. In this sense it is linked to a universal aspiration, not a particular culture. The counter-position of community and reciprocity to travel can express a profound conservatism. One commentator even advocates 'a policy of moderate Nimbyism' (Not in my back yard-ism) as 'the only way of preserving our healing contact with nature'.[65] Insularity and parochialism can in this way become virtues in the critique of Mass Tourism.

The tourist experience

For the New Moral Tourist, host and guest are defined by difference, the negotiation of which is a fraught affair. But in reality the host-tourist encounter is less problematic than Poon's New Tourist and much of the anthropological literature suggests. The tourist encounter may take place in the context of inequality but it is not the cause. It brings together, face to face, peoples divided by unequal access to society's resources and opportunities. The New Moral Tourist collapses the unequal relationships that exist between countries into simply being a tourist, and comes to see interpersonal conduct itself as part of the problem, as well as the solution. How we behave and consume becomes the subject of debate, rather than the unequal *context* in which we relate to one another. A large dose of personal guilt for the tourist is the result; a guilt that can apparently only be assuaged through 'ethical tourism' practices.

The perceived need to negotiate cultural difference has fuelled a fraught, angst-ridden debate. It is conceivable that the trend towards codes of conduct and guarded behaviour implicit in New Moral Tourism can denigrate the tourist experience itself. Young, independent travellers, striking out and learning for themselves, are confronted with a plethora of well-meaning advice and codes of conduct amounting to an attempt at moral regulation. Some environmentalists even suggest it would be better to avoid the risks and holiday closer to home if at all.[66] Yet an important aspect of tourism is the sense of personal autonomy and freedom – there is much to be said for trusting oneself and learning from one's own mistakes in the transition from youth to adulthood, a transition in which leisure travel often plays a role.[67] Moreover, the assumption of the primacy of difference colours the way we see the host. Conditions of relative poverty can be rationalised, and even celebrated, in cultural terms.

A starker summary of the argument is evident in the following anecdote gleaned from an acquaintance. A young St Lucian waitress dresses in traditional costume because this is how she thinks the tourist has envisioned their host. The tourist asks earnestly about the local way of life, because this is the sort of thing a thinking tourist should do. In the context of such role-playing, it is very difficult to discover empathy or commonality. The student traveller concerned was studying at university. The young woman was saving hard in order to travel, and hopefully study abroad. Only after a few drinks, and a loosening of cultural mores did they uncover their common aspirations. Material inequality between North and South, not a fetishised 'culture', set the two apart. The thrill of travel is that, one individual to another, barriers can be stripped away and real friendships made. The New Moral Tourism seems intent on erecting new barriers based on a notion of a decentred, self-limiting tourist, who is too busy gazing at the Other's culture to empathise with them as an individual. Ultimately, the conclusion of the anthropological outlook is that if you look for Otherness, you will find it. You can wonder at it, and wonderful it often is. But an oversensitivity to cultures borne of a contemplative tourist gaze can equally be a barrier to unrestrained, open communication.

Conclusion

The conception of culture adopted is not only an aesthetic concern. New Moral Tourism is not just a romantic view of the world that informs leisure choices – it is a moral crusade. How the host is viewed, through the prism of culture, inevitably affects the prospects for development and type of development on offer. Culture defined as *function, the past* and *difference* effectively creates culture as a straightjacket for societies that may desire economic development. As I examine in chapter 7, this can have a bearing on development policies themselves. Culture becomes objectified; a romantic image cast in stone, rather than the creative subjectivity of the host. It can become a part of heritage, the past, preserved for the sensibilities of the tourist, rather than being made and remade in the context of social change.

Of course tourism need not carry some of the problematic assumptions outlined here. Interest in the past does not necessarily imply turning the host into a museum piece. The reality in economically developed regions is that distinct cultures survive in the modern world. Hybrid cultural forms embody change through taking what is valued from the past and what is desired for the future. More crucially, cultural change is less likely to be seen as destructive and more likely to be seen as creative in more developed societies.

Tourism can lead to wider contact between peoples from different regions and cultures. Whilst some see risks in terms of 'cultural imperialism',[68] it can also be the case that contact between people breaks down any preconceived notions of Otherness. It is not unusual to experience 'culture shock' in reverse – we expect a world of difference but discover people who aspire to the same things as we do, be it within quite different situations. Part of the thrill of travel is to meet and connect with people outside of one's own community or culture. But to view the host as a cultural icon, representative of a society unsullied by modern life, is to diminish the potential in the tourist experience for both host and tourist. Oversensitivity to Otherness blinds us to common humanity.

The important point here is that it is easy to see culture rather than *people*. Whilst notions of culture, in the context of tourism, often carry romantic assumptions about ways of life unsullied by consumerism, the reality is often somewhat more straightforward. Culture, for those living in poorer countries, often corresponds more closely to the original meaning of the word – tending crops and animals – a way of life dictated by living on the margins of the global economy rather than a cultural lifestyle choice.

A street in Pokhara, Nepal. Pokhara is a popular base for
trekkers in the Annapurna region. (Photo: Karen Thomas)

6 Travelling for a change

Global culture and the ethical tourist

Tourism is viewed as a destroyer of cultural diversity by many of its critics. This chapter argues that this *cultural* critique of modern tourism misses the *social* context of the poverty and inequality that many tourists come across on their travels. New Moral Tourists seek to make a difference to the problems they perceive through what and how they consume – a mixture of ethical consumption and lifestyle politics. Yet this approach ends up in empty moralising about individual choices and fails to address the causes of substantial inequalities between societies.

Globalising tourism

Globalisation has caught the mood of today. Protesters proclaim their anti-globalisation stance prominently and authors rail against the levelling of cultures around a western standard. Tourism is often in the frame for contributing to a globalisation of culture. For one author, tourists 'seem to be the incarnation of the materialism, philistinism and cultural homogenisation that is sweeping all before it in a converging world'.[1] According to this view tourists are carriers of the modern disease of materialism, and there is an imminent danger that they may infect others on their travels.

Tourism Concern's *Community Tourism Guide* is worth quoting at length on this, as it epitomises the critical take on tourism-as-globalisation:

> One day, somewhere deep in the rainforest in South America or Borneo or Central Africa, a few nervous men and women will step into a muddy clearing in the jungle. Cautiously, they will accept the steel machetes or cooking pots being held out by a government-sponsored anthropologist, before hurrying back into the safety of the forest.
>
> The encounter will not be marked by any great fanfare. It will probably not make the news. Yet it will be a significant landmark in human history. The last 'uncontacted' tribe on earth will have been caught in our global web, and an era of exploration, invasion and global integration that began when Columbus first set eyes on the Americas will be

over. For the first time, the entire human race will be connected in one giant, all-embracing cultural and trading network.

As this era of human history comes to a close, we are left with a dominant social and economic system that ignores human and environmental costs. A system that destroys communal life because of its demand for a mobile labour force. That creates mental illnesses and stress by sucking people into huge, anonymous cities. That discourages people from growing their own food because doing so doesn't involve selling anything (and therefore doesn't show up as profit in economic statistics). A system that puts a greater value on a pile of dead wood than a living forest.[2]

Of course, globalisation is not a new phenomenon – capitalist development has always been global by its nature. However, it is commonly held that 'cultural convergence' is a relatively recent, and qualitatively new, phenomenon.[3] Terms such as 'cultural levelling', or the 'Americanisation of culture', are often invoked as the destroyers of cultural diversity, the latter held in esteem.

The view that cultural difference is ignored in development more generally has wide resonance. Many environmentalists, in similar vein to Mark Mann, author of the above extract, argue that western societies are engaged in imposing their own values on to other peoples in lesser developed parts of the world. Philip MacMicheal, for example, talks of 'western developmentalism' rather than development. This 'western developmentalism', he argues, takes its cue from an 'economic model driven by technology and market behaviour rather than from existing cultures'.[4] Modern technology and markets are in this formulation distinctly 'western', and MacMicheal argues that other societies should be able to develop in ways more in keeping with their own culture.

This is a sentiment widely shared by the advocates of New Moral Tourism, who propose development in keeping with the way rural Third World societies function – culturally sensitive development is central to the moralisation of tourism. Such an approach, it is claimed, does not impose development, but consults with communities and works with *what is* rather than advocating transformative economic change.

The advocacy of culturally appropriate development is especially prominent in debates on tourism because tourism not only involves an economic relationship between the producer and consumer, but also a direct cultural one between consumers and their 'hosts' – effectively tourists travel to consume their experience and the services they have paid for. In this respect tourism is quite different from other industries. Three authors introducing a study of gender and tourism explain it thus: 'tourism is consumed at the point of "existence" and involves more than the material; it is cultural; it involves gazing upon and the "selling" of "otherness" and the unique'.[5]

It is this aspect of the international leisure travel industry that has made it prominent, along with television and fast food, in the list of those

accused of cultural globalisation – it is held to carry the potential to undermine the cultures of host communities through direct contact with the
culture of the visitors. Deborah McClaren of the Rethinking Tourism Project
in the USA, and author of *Rethinking Tourism and Ecotravel: the Paving
of Paradise and What You can Do To Stop It*, puts the global threat to local
cultures thus:

> Global tourism threatens indigenous knowledge and intellectual prop-
> erty rights, our technologies, religions, sacred sites, social structures
> and relationships, wildlife, ecosystems, economies and basic rights to
> informed understanding: reducing indigenous peoples to simply another
> consumer product that is quickly becoming exhaustible.[6]

Elsewhere, McClaren and co-author Lee Pera fear that as a result of cultural
products being included in the World Trade Organisation's policy of
reducing barriers to trade, a 'tourist monoculture' characterised by consumerism is theatening to overwhelm indigenous cultures. The authors
uphold the innate value of indigenous knowledge, which they feel 'must
be safeguarded'.[7]

One problem with such opposition to tourism's globalising tendencies is
that in arguing against the extension of global culture (typically railed against
in the form of McDonald's, Coca-Cola and Nike ... and in this context,
Mass Tourism), in favour of cultural diversity, critics are reluctant to argue
for material equality and the growth needed to bring it about. Aspirations
for a better life, including the aspiration to travel for leisure, are surely examples of a common human culture, a global culture, and one that should
be celebrated rather than condemned. In this sense, one could argue that
global culture is precisely what needs to be extended in the name of
promoting equality.

Globalisation, in its cultural usage, suggests combined development – one
size fits all – normally in the form of 'Americanisation' or 'westernisation'.
It is this that many of Mass Tourism's critics find so objectionable. But capitalist development has historically been characterised not only by its
combined, but also by its *uneven* nature. Growth in some parts of the world
has tended to rely on a dominant economic relationship with other parts of
the world – the development of the former has shaped the development, or
lack of it, of the latter. Whilst critics of globalisation tend to criticise what
they see as a market-driven imposition of cultural sameness, they are less
vocal about the market-driven denial of material equality. The explicit agenda
of New Moral Tourism is a defence of the host society in the face of the
culture of the tourist and the tourism industry. However, reflected in the
promotion of cultural diversity and environmental conservation is an implicit
acceptance of the status quo with regard to the broader inequalities that
characterise the world we live in.

The tourist and cultural domination

Perhaps the most striking criticism of Mass Tourism is the charge that it constitutes cultural domination, or even colonialism or imperialism, in the Third World. This is not a rare criticism. Jost Krippendorf, in his seminal *The Holiday Makers*, asserts that tourism has a colonial character 'everywhere and without exception'.[8] The influential *The Golden Hordes: International Tourism and the Pleasure Periphery* by Turner and Ash regards it as a 'new form of colonialism'.[9] In similar vein one campaigning non-governmental organisation (NGO) prominently beg the question 'Tourism – the New Imperialism?' in their literature, and proceed to answer this with a litany of charges against 'unethical' tourism.[10] Anthropologist Dennison Nash is associated with this perspective, which is widely shared amongst campaigning groups and writers critical of Mass Tourism.[11] Nash refers to imperialism as the expansion of a society's interests abroad, whether they be 'economic, political, military, religious or some other'.[12] Such a broad conception of imperialism effectively takes in any expression of inequality between societies and it is in this context that tourism can become 'imperialism'. Formerly understood to refer to the systematic division of the world into wealthy countries and those plundered to benefit the wealthy, imperialism has been re-presented as an unequal relationship between individuals, in this case the tourist and the host. It is no surprise, then, that the tourist to the Third World is viewed in a circumspect fashion. They are seen as embodying, even personifying, the unequal relationships between countries and regions. Hence imperialism is presented as the injustices of individuals, along with nations and companies, against others. In this analysis tourists become part of a chain of domination, along with tour operators, hotel chains, banks, and presumably national armies – the tourist becomes an imperialist, implicated in the subordination of the host society. The result of this is that Nash can cite the North American traveller who wants fast food and hot water as part of a systematic domination of the Third World.[13] One prominent academic even suggests a continuity between today's tourists and invading armies of previous eras: 'The easy-going tourist of our era might well complete the work of his predecessors, also travellers from the west – the conqueror and colonialist.'[14] These pronouncements surely reflect a relativisation of domination and colonialism that makes the words almost meaningless.

Seeing the world in this way involves conflating relations between individuals and broader power relationships between countries and classes. This is made explicit by one book, which argues that, 'Tourism . . . involves the purchase of the particular social relations and characteristics of the host',[15] and that people, host and tourist, embody 'social relations'. This suggests that tourists are complicit in, or even embody, the relationships of inequality that characterise the world. The logic of this is that they should alter the way they relate to others to alleviate these problems.

Implicit in the conflation of individual relationships and social relation-ships is the theory of commodity chains.[16] Put simply, this holds that when a person consumes, they enter into a global process that has implications for the whole of society. There is a chain binding together consumption with production – some people buy it, so others produce it, this produc-tion creates subsequent demands bringing forth supply in other areas of the economy. Commodity chains is a useful idea for conceptualising how the economy works. However, it takes a conception of society as the sum of its parts, and hence it suggests that to a greater or lesser degree we are all complicit in the problems faced by Third World countries, as we are all part of a continuum, the commodity chain. It implicitly points to our role as consumers as key to addressing these problems. The New Moral Tourism, then, sees people's moral role in terms of the role of individuals at the level of consumption. Even if we were all to agree on the problems to be tackled, this is a very limited arena for moral action.

The complicity of the tourist in reinforcing inequality, both cultural and economic, is deemed to be a consequence of a lack of awareness. The tourist is deemed unaware and in need of enlightenment regarding how they act on holiday. They are, as leisure travellers, reproached and encouraged to take on board a degree of responsibility for their hosts. This is the message of the plethora of codes of conduct, ethical tourism guides and New Moral Tourism tour operators, too. Of course, the emphasis on changing tourists' behaviour does not necessarily imply a wilful desire on the part of holiday-makers to be complicit in cultural domination. Rather, their impacts are more likely to be seen as an unconscious cultural bias. In this vein Polly Pattullo blames the lack of linkages of tourism investments in the Caribbean with small local business on the 'entrenched conservatism of the package tourist' who want familiar surroundings and western food.[17] Such tourists, she argues, contribute to a subordinate cultural and economic position for their hosts, not through a conscious bias against other people, but through 'conservatism'.

The new forms of tourism are designed to counter this – they are seen as offering a break with the patterns of domination of the past. In this way New Moral Tourism has been presented as an antidote to the cultural arro-gance of Mass Tourism. Industry analyst Ahluwalia Poon has even suggested that the marketing of a country around 'new tourism' themes and products is one means of countering foreign domination embodied in the big multi-national chains and also in the stereotyped portrayal of the Third World in holiday brochures.[18]

Such an approach is presented as a radical critique. Yet in focusing on culture, taken in this context to mean beliefs embodied in consumption decisions and individual behaviour, the critique is in fact a conservative com-ment on people, both host and guest. It assumes that individuals are domi-nated by other individuals; that their culture is destroyed through a 'demon-stration effect'. Commentaries on tourism often refer to this demonstration effect: the assimilation of western, or commonly American, tastes and consumption patterns by other parts of the world. It is often seen in a negative light – outside influences from wealthy tourists are viewed as diluting local cultures, perhaps creating expectations amongst younger members of a society that cannot be met, leading to inter-generational tensions.

Yet the fact that poor people may identify with symbols of affluence signifies that they may want some of this affluence. That it is out of reach is an argument for economic development rather than protection from global culture. The New Moral Tourism's opposition to cultural 'imperialism' does not argue that other countries should have access to hotels and leisure travel and the good things in life, but that today's tourists should 'leave well alone' on their holidays.

Far from being very radical, this approach embodies some very conservative assumptions about host societies – notably that aspiration and potential to improve these societies through development take second place to a sense that they are victims of the tourists' culture. In fact it could be that the critics of the demonstration effect are doing the very thing they accuse western governments and companies of being responsible for – superimposing their own version of development, in this case one that is intensely critical of the modern project, on to societies crying out for the material benefits of modern societies.[19]

Geographical inequalities?

One outcome of viewing inequalities in terms of culture is that they readily become presented and understood in geographical terms; it is held that there are tourism-generating countries reflecting western, more affluent cultures, and there is a 'pleasure periphery' which is located in the Third World and in economically peripheral regions within the developed world, reflecting a rich cultural diversity. As a description, the notion of the pleasure periphery is quite limited. It is simply not the case that the majority of tourists head for poorer countries on their holidays. The large majority of tourism is from developed countries to developed countries – hardly surprising perhaps, except in the context of a discussion that presents the developing world as the primary victim in a rapacious drive for leisure travel.

Of course it is true that gross inequalities are evident between different countries and regions of the world. But a discussion of divisions and inequality should recognise that these divisions exist sociologically as well as geographically. For example, social class is rarely invoked in the moralised discussion of tourism. To put the point succinctly, do hotel staff in Miami have more in common with hotel staff in Nairobi, or with wealthy American tourists travelling to Kenya? And if the Miami hotel receptionist saves enough for the holiday of a lifetime in Kenya do they join the club of potential 'cultural imperialists', potential abusers of the host society? Do they become part of the guilty North? Advocates of New Moral Tourism habitually castigate Mass Tourists in the context of criticising tourism as cultural arrogance or even domination. An irony here is that the masses are presented as the new colonialists, whilst governments overseeing debt and meagre aid budgets counsel the masses on their aggressive, imperialistic attitudes through their support for the rhetoric of ethical travel.

From consumption as problem to consumption as solution

The thrust of New Moral Tourism is that mass consumption of package holidays has the potential to level cultures and contribute to a global uniformity that diminishes us all, tourist and host. Further, it is held that tourists should change their behaviour and be more thoughtful and conscious of their individual contribution to perceived global problems. It is asserted that they should modify their attitudes towards their hosts in such a way as to encourage diversity and conservation. This is essentially the politics of lifestyle – trying to achieve the dual aims of protecting cultural diversity and conservation of certain environments through one's conspicuous lifestyle choices in the realm of consumption (holidays have long been the subject of dinner party conversation, but today one can expect such conversations to extend to one's ethical credentials). Hence the corollary of seeing tourism as cultural globalisation in this way is to advocate ethical conduct in the forms of ethical consumption and ethical behaviour.

Ethical consumption

New Moral Tourism is a form of ethical consumption, and as such is part of a wider agenda that has become established over the last twenty years. The notion that people try to make a difference to the world in which they live through what they buy and where they buy has become a commonplace part of contemporary political culture.

In 1981 *The Green Consumer* sold some 350 thousand copies in a single year.[20] It was illustrative of the growth of ethical consumption as a focus for people's aspirations for social change (or perhaps more accurately to slow down change). Of course, engaging in social action through consumption may be social action only for those who can afford to pay, and in the context of the New Moral Tourism this may be apposite – often buying the new, ethical brands involves paying a premium. However, there is much evidence to suggest that the ethical consumer agenda has a wide resonance. Numerous surveys show that large numbers of people view themselves as green consumers, and furthermore, that green, ethical consumption cannot be dismissed as the prerogative of salaried middle-class sentiment.[21] Rather, it is a pervasive agenda.

Put simply, it is argued that consumers can force a more ethical agenda on to companies through exercising choice in favour of products that are deemed more sustainable. Such a view is personified by The Body Shop founder Anita Roddick. 'Don't just grin and bear it. As consumers we have real power to affect change . . . we can use our ultimate power, voting with our feet and wallets – in buying a product somewhere else, or not buying it at all.'[22] In fact Anita Roddick has of late also become involved in the ethical holiday agenda through her involvement in the lottery-funded ethical tourism magazine *Being There*, distributed through The Body Shop. Many of the campaigns, NGOs and web sites promoting New Moral Tourism advocate ethical consumption along similar lines.[23]

The growth of ethical consumption reflects important shifts in the way people relate to society more broadly – the issues they prioritise and the ways in which they may seek to have an impact upon these issues. How we consume has grown in prominence relative to the workplace as the terrain on which identities are formed and social issues are debated. Sociologist Zygmunt Bauman puts it thus:

> in present-day society, consumer conduct (consumer freedom geared to the consumer market) moves steadily into the position of, simultane-ously, the cognitive and moral focus of life, the integrative bond of the society . . . In other words, it moves into the self same position which in the past – during the 'modern' phase of capitalist society – was occu-pied by work.[24]

Bauman makes the widely accepted point that in previous periods, the realm of production, or work, was more central to identity, but that today it is more as consumers that we develop a sense of ourselves in the world. The growth of the importance of consumption is often viewed in very positive terms. The world of consumption and identities, personal and political, takes the appearance of a world of choice and freedom – one can break free of traditional collective identities connected to class, race or gender and develop one's own identity and experiment in ways not evident in the past. In polit-ical terms, too, it has been argued that consumers can generate pressure for change in a way traditional, discredited political institutions are unable to.[25]

Many argue that this shift is in part due to changes in the economy and the nature of work. The argument that post-Fordism – a shift towards a more individuated, less collective experience in the workplace – contributed to a decline in traditional collective allegiances relating to work was developed in the pages of the erstwhile house journal of the left *Marxism Today* by Stuart Hall, Charles Leadbetter and others.[26] Post-Fordism was notable in that it marked a shift from the politics of production (and social class) to con-sumption (and individual identity) in radical thinking. The relative shift towards consumption is very much part of New Moral Tourism – consumption of tourism has become in a sense politicised, or more accurately, moralised.

However, more profoundly influencing this shift is the collapse of perceived alternatives to capitalism. The collapse of communism seems to confirm that alternatives to the market do not work. This is reinforced by the adoption, or at the very least acceptance of 'market forces' as positive or in-effaceable even by capitalism's erstwhile critics on the Left. This has contributed to a lack of questioning of the market, which has taken on the appearance of an eternal reality in political and social debates. Francis Fukuyama's *End of History* thesis, following soon after the end of the Cold War, presenting a contemporary world in which all the big ideological issues have been settled, is perhaps emblematic of a sense of closure of grand politics.[27]

The decline of allegiance to big political ideas has contributed to a disconnection between individuals and their governments, and has led to a preoccupation with re-establishing this connection in some way. However, traditional political channels increasingly invite cynicism, and many feel alienated from the institutions of government. Other institutions, through which individuals related to their society have also declined – church, community and family. All this has strengthened, by default, the more *individual* form of politics – consumer politics. Far from the discredited institutions of government, it is as consumers that we are, apparently, free to exercise our choice in pursuit of a better world.

In short, the 'consumer-producer' relationship has grown in prominence as 'worker-employer', the politics of social class (a politics that has something to say about the terms on which production is organised) have declined in influence. This process, which can only be outlined here, elevates the importance of the world of consumption in the search for selfhood. How we see ourselves in relation to others, and in relation to society more broadly, seems to increasingly take consumption as its primary point of reference.

This trend is reflected in domestic politics in the UK and elsewhere, which have witnessed a reorientation of Left and Right around a pragmatic centre, a centre in which issues are increasingly reinterpreted in terms of consumption. Issues such as the health service, rail and public services, once discussed in terms of their part in a socialist or collective imperative, or alternatively questioned as a drain on resources that could be employed more productively through the market, are today viewed through the prism of the consumer – 'meeting customer needs' has become the mantra of public and private sector alike. Left and Right, and their corollary economic theories, Keynesianism and monetarism, were about, in part at least, how production should be organised and the role of the state in this. Today the emphasis is on how to meet the needs of consumers of public services. Even the trade unions' dispute with employers – which has traditionally been very much about the contestation of the terms on which production takes place – is most commonly conducted through a discourse about how best to meet the needs of paying customers, with employers claiming they have to increase efficiency by controlling costs, and employees arguing that safety standards will fall if proper pay and conditions are not forthcoming.

The corollary of the decline of traditional politics, connected to different and distinctive views on the ways societies should organise production, is the rise of consumer politics. This broader social perspective shapes the debates on tourism and travel. However, because how, where and what we consume is essentially an individual matter, a matter of behaviour, it is better described as a moral dimension than a political one. Buying a holiday, and how one conducts oneself abroad has, for some, become a conspicuous expression of morality.

So the closure of the politics of the contestation of production, the politics of class, has contributed to a reorientation of social action towards the far more limited realm of the private consumer. The decline of 'big politics' is in part a result of the disillusionment with the traditional, formal political process and parties. Across the world, political allegiances seem to a greater of lesser degree to have crumbled over the last decade and a half. Consumer politics stands apart from this process – it involves no allegiance to grand schemas, no association with apparently failed political projects and eschews collectivity in favour of individualism. This shift in where one can 'make a difference', and also its limited horizons, are well expressed in the following:

> the past decade has witnessed a massive loss of confidence in what may be held to be the bedrock of formal democracy. Faith in government, in the credibility of politicians, in the power on governments to do anything, has hit an all time low . . .
> Is there really nowhere to go but the shops?[28]

The shops, or perhaps in this case the independent travel agent, is a place to make a difference or say something about oneself in the wider world. The author goes on to make a case for 'the shops':

> What needs saying at this stage is that our conception of politics must be prised open . . . Today's consumer culture straddles public and private space, creating blurred areas in between. Privatised car culture, with its collective Red Nose days and stickers for lead-free petrol; cosmetics as the quintessential expression of consumer choice now carry anxieties over eco politics. These are the localised points where consumption meshes with social demands and aspirations. So the above cannot be about individualism versus collectivism, but about articulating the two in a new relation that can form the basis for a future common sense.[29]

Whilst this author claims that ethical consumption is not an individualist agenda counter to collectivity of the past, its collectivist credentials are questionable. Consuming is what we do individually, whilst in production people are more likely to be part of a collective, defined by place of work or a broader sense of class interest. Groups of individuals may be similarly motivated to buy certain types of products, but even if one was to agree with their outlook, their role as consumers makes challenging the basis on which society produces all but impossible. Yet it is this basis that informs the problems of poverty and lack of development. For example, it is the unequal economic relationships between *countries* that result in the net transfer of material wealth from the Third World to the developed world alongside the large monetary indebtedness of the poorer nations. But it is these very relationships – relationships between nations rather than individuals, based on the drive for profit – that are presented as beyond the realm of human action due to the dominance of global markets.

The idea of globalisation seems to reflect these limited horizons. Globalisation holds that what someone buys in one country has effects globally, in both an economic and cultural sense, and hence suggests that what we buy can make a positive as well as a negative impact. Sociologist Anthony Giddens, in his book *Beyond Left and Right*, explains the link between globalisation and ethical consumption:

> Our day-to-day activities are increasingly influenced by events happening on the other side of the world. Conversely, local lifestyle habits have become globally consequential. Thus my decision to buy a certain item of clothing has implications not only for the international division of labour, but for the Earth's ecosystem.[30]

In this way, globalisation is often invoked to emphasise the interconnected nature of society – we are all bound together through the market. But globalisation often carries the underlying implication that 'the market is beyond human intervention'.[31] Hence whilst we are encouraged to see ourselves as ethical in our role as consumers, the basis on which we consume, the power relationships between nations and between social classes, appears beyond us. Ethical consumption hence reflects a very limited moral universe, as it is one that shies away from challenging the notion that there is life, and politics, beyond the shops. However one views them, societies' problems cannot be addressed from the basis of consumption – it is an illusory realm of social action, and as such an illusory realm of morality too.

Lifestyle Politics

Another way of looking at New Moral Tourism, closely related to ethical consumption, is as a form of Lifestyle Politics – political solutions at the level of the individual in their daily lives. This is the level at which New Moral Tourism offers solutions – what one buys and how one relates to other people.

The problems for which solutions are offered are generally environmental ones, and hence individual lifestyle is very much part of the debate on how to move towards sustainable lifestyles. For example, *Caring For The Earth*, a report published by the International Union for the Conservation of Nature in 1991, argues that resource problems are *human* problems rather than distinctly environmental ones.[32] This is developed in the report, which puts an emphasis on the need for more sustainable lifestyles within the developed world. Reflecting this, how we act and what we consume have become more central in the debate on sustainability. Consumption and lifestyle have become central to the advocacy of sustainable tourism, too. Moreover, this argument is especially pertinent to the tourism industry given the way that tourism can be seen as conspicuous consumption of the more wealthy nations, and that tourism tastes are strongly associated with image and lifestyle. A moral dimension to lifestyle is central to the moralisation of tourism.

In this vein the World Wide Fund For Nature (WWF), a conservation NGO that supports ecotourism as a way of delivering conservation and alleviating rural poverty in the Third World, argues that, 'Ecotourism is not a product, but an attitude.'[33] New Moral Tourism is an attitude to life, and how one conducts oneself, linked by this organisation and other advocates of ecotourism to benefiting the environment. Poon, analyst and advocate of New Tourism, reflecting the crusading character of those with this 'attitude' goes so far as to say that ecotourism must become a 'way of life'.[34]

Anthony Giddens talks of a shift from the traditional politics of emancipation, embodied in traditional collective ideas such as trades unions, towards life politics.[35] Life politics refers to individuals' attempts to reposition themselves culturally in the context of their own lives and through this to try to make a difference to their immediate environment and also more broadly.

Giddens argues that life politics is less a retreat from the social world into the individual, and more a reconfiguration of the relationship of the individual to their society – identity becomes a site of political change. Whilst some see the growth of the importance of life politics, or indeed the politics of consumption, as a reconfiguration of the social, others point out that its importance has been parallel to the *decline* of the social.[36] In other words, lifestyle represents a narrowing of human subjectivity away from collective solutions towards individual ones; away from broader social relationships towards those between individuals. As people are less likely to view traditional political channels as so central to their lives, their beliefs and aspirations, it is conceivable that a lack of engagement with society through these channels may, by default, elevate a sense that one's lifestyle is a key arena in which to express one's beliefs and make a difference.

Dean MacCannell

The politics of lifestyle places importance on to how people relate to one another in their daily lives, and in this instance on how they relate to people with different cultures in different places. As such, it problematises the relationships between individuals.

The shift towards regarding relationships between people, and between people and places, as culturally fraught is well illustrated in Dean

MacCannell's sociology. MacCannell, in his book *Empty Meeting Grounds: the Tourist Papers*, usefully sets out his stall as being someone wedded to the politics of social class, but who has seen the prospects of progressive change through the working class diminish. So whilst he recognises the decline of human subjectivity in the form of the working class, he sees cultural encounters, such as between host and tourist, as having the potential to lead to the formation of 'new subjects'.[37] MacCannell hence turns to the cultural realm, and even the interpersonal realm, looking for ways to make the world a better, more human and humane place to live.

MacCannell's project is laudable, but also seems very limited. Leisure travel becomes an aspect of life imbued with the potential for progressive change – a sort of politics of everyday life. Indeed, this is precisely what New Moral Tourism represents – the channelling of desire for change into something as everyday as leisure pursuits.

However, MacCannell's view represents a retreat into the politics of the personal. The decline of class as a vehicle for change has contributed to a profound cynicism with politics and a pervasive anti-political mood in society. MacCannell's cultural politics rationalises this retreat and tries to draw something positive from it – if we can't change the world, at least we can change the way we relate to other cultures and societies. If we have little power in the workplace, we can 'rehumanise' human relations through how we relate to each other, even as tourists.

Moreover, it is informed by a *fin de siècle* rejection of human progress in favour of deference to nature and cultural difference. In one section of *Empty Meeting Grounds*, MacCannell rails against what he sees as the imposition of science on to poor, rural inhabitants of the Third World. He observes that these people enjoy their 'primitive' life, and suggests that the developed 'Anglo-Europeans . . . are often absolutely intolerant of the joy of others'.[38] MacCannell's association of joy with extreme poverty is characteristic of his dim view of the material gains made in modern societies, and the sense that poorer societies may lose out from the cultural encounters between themselves and tourists.

MacCannell expresses the basis for the moralisation of tourism – the sense that inter-cultural and interpersonal relations is the terrain of social change. However, the appearance of choice here masks the narrowing of real choice – the emergence of the cultural subject, free to choose and to act in the realm of lifestyle, marks the narrowing of political contestation of how society is organised. It marks a degraded form of subjectivity, incredibly limited in scope, and often reactionary in outlook.

Conclusion

The result of the emphasis on interpersonal morality is guilt and angst for tourists. In fact, the individual solutions to social inequalities proposed by the New Moral Tourism lobby are deceitful. They effectively take people's aspirations to do good and convert them into personal guilt at the poverty evident in Third World destinations. Yet to travel or not to travel, to stay in a hotel or in a village, to enjoy the culture or just the climate, will make no difference to the broader inequality that exists between nations and peoples. More importantly, it is an agenda that discourages a critical examination of the causes of poverty by presenting individual behaviour as a strategy to bring positive change to the Third World. This makes for degraded politics, and a diminished travel experience too.

A sign in Hwange National Park, Zimbabwe.

Zimbabwe has a thriving elephant population, numbered at over 64,000, up from 4,000 in 1900. The non-governmental organisation funded 'Campfire' scheme supports the elephant population through ecotourism and organised hunting in communal lands, thus giving the elephants an economic value and creating an incentive for Zimbabweans to get involved in conservation. The success of Campfire is seen as turning elephants from 'foe to friend' in communual areas, which often buffer national parks and hence are prone to crop damage or even threats to life from marauding elephants from the parks. It is estimated that around 250,000 Zimbabweans are involved in managing their natural resources through the Campfire scheme. (Photo: Cheryl Mvula)

7 New Moral Tourism, the Third World and development

The claims of certain forms of tourism to be more moral are not only rooted in the ideological developments referred to in previous chapters. They are also reflected in the conception of nature-based, culturally sensitive tourism as an exemplary development tool, especially in the Third World. Mass Tourism's impact on development is deemed to be disappointing and highly problematic from a cultural and environmental perspective. It is considered that New Moral Tourism types, however, such as ecotourism and community tourism, are able to bring together the previously antagonistic goals of development and conservation.

This chapter examines the claims made for New Moral Tourism in the field of development.

New Moral Tourism as development

It is widely argued that whilst Mass Tourism has brought problems in its wake, new types of tourism such as ecotourism and community tourism are beneficial with regard to development in the Third World. The claims that these types of tourism are ethical are based on their professed capacity to meet conservation aims whilst providing benefits for communities. Ecotourism, community tourism, nature tourism and so on are not only part of the etiquette of the New Moral Tourism movement but have become the focus of various NGOs and campaigns, new and old, to achieve this goal, especially in the world's poorer countries. They attract considerable aid funding through governmental and supragovernmental aid organisations too.

The tourism-for-conservation agenda is an important one. *The Green Travel Guide* argues that, 'tourism can be a powerful force for conservation' and notes that there are more than 5,000 national parks, wildlife sanctuaries and reserves around the world today, many depending on tourism for financial support.[1] USAID, the aid arm of the United States government, for example, use ecotourism as a strategic tool for 'environmentally responsible development' in more than a dozen countries.[2] Conservation International, a big and wealthy player in international conservation, utilise it in seventeen out of the twenty-five countries in which they operate.[3] In the UK the

Department for International Development run a scheme promoting what they refer to as 'pro-poor tourism', aiming to help the very poorest in rural parts of the Third World by attracting tourists appreciative of the undisturbed environment.

Such an approach is now commonplace within the NGO and aid world. NGOs, scientific organisations and conservation organisations, such as the WWF, Nature Conservancy, the Audubon Society, the Sierra Club and the Earthwatch Institute promote ecotourism for similar reasons. It extends to smaller campaigns and NGOs, too. The US-based International Eco-Tourism Society propose ecotourism as a boon to Third World countries threatened by what they perceive to be harmful development. Ecotourism, they believe, can achieve both conservation and development, two goals often considered to be antithetical.[4] Similarly, influential campaign Tourism Concern see community tourism as an ethical alternative to what they regard as damaging Mass Tourism. Table 1 shows a selection of projects concerned with conservation and development through ecotourism.[5]

The conservation-oriented NGOs especially have tended to benefit from what has been termed the 'greening of aid' – the tendency to attach environmental conditions or emphasis to Third World aid. They are also considered characteristic of a growing 'civil society', having grown greatly in number and in influence. They have acted as powerful vehicles for the transference of environmentalist thinking into the arena of development policy, and it is here that nature-based tourism is considered innovative.

Tourism and environment – symbiosis?

How can tourism, castigated as a problem, and even as imperialism, in its 'mass' form, become advocated as a solution in its 'new' forms?

One author answers this succinctly in a paper titled 'Tourism and Natural Heritage, a Symbiotic Relationship?'.[6] Harold Goodwin argues against the view that tourism, like any other industry, is likely to be in conflict with the natural environment. Rather, he argues, it has a special role to play in development. Nature tourism, depending as it does on a desire to experience areas of perceived natural beauty and distinction, can provide funds to manage and maintain these areas. Consumers may be prepared to pay considerable amounts of money in order to experience 'untouched' environments. This money can then be used, in part, to encourage people local to the area to co-operate in the conservation of their environment.

This perspective is important as it questions, as Goodwin points out, the perception that there is necessarily a tension between development and the environment.[8] Put simply, some may see industrial development as vital for creating jobs and wealth, whilst others may point to the environmental impact of such development. Prioritising one means neglecting the other – a 'win–lose' scenario. Tourism, it is argued, is one form of development that can go some way to resolving this tension – if it is the right kind of tourism,

Table 1 Examples of projects supporting ecotourism as a means towards integrated conservation and development[5]

Country	Name of project	Donors involved	Special features
Belize	Rio Brave Conservation and Management Area	various American and British donors, private sponsors	education and research-related ecotourism successfully managed by an NGO (Programme for Belize)
Brazil	Proecotour	InterAmerican Development Bank (IDB)	large-scale ecotourism programme covering the entire Amazon, financing infrastructure, private investments, marketing studies, private sector capacity building (in preparation)
Honduras	Ecotourism development in protected area system	UN Development Programme (UNDP)/ Global Environment Facility (GEF), World Bank	area selection according to tourism potential; ecotourism facilities, training and promotion, NGO capacity building
Central America	Promotion of sustainable development through tourism (FODESTUR)	German Agency for Technical Cooperation (GTZ)	facilitation of regional stakeholder co-operation and creation of regional ecotourism routes (Ruta Verde)
Cameroon	National Ecotourism Strategy	German Agency for Technical Cooperation (GTZ)	national ecotourism strategy focussing on stakeholder participation, protected areas and local communities (in preparation)
Namibia	Namibian Community Based Tourism Association (NACOBTA)	Swedish International Development Cooperation Agency (SIDA), Worldwide Fund for Nature (WWF)/US Agency for International Development (USAID), various European donors	community tourism development based on Communal Area Conservancies; community funds, training and marketing
Tanzania	Cultural Tourism Programme	Dutch Development Organisation (SNV)	development of ecotourism products in indigenous communities near protected areas, capacity building and institutional strengthening; marketing through National Tourist Board

Holiday snaps – tourism: the lesser of development evils[7]

(from questions and answers on ecotravel, from US conservationists the Audubon Society's 'The Ethics of Ecotravel')

Q: There's a boom in ecotravel and adventure travel. Should we visit wild places, or should we leave them alone?

A: There are fragile environments, such as areas of the Amazon, Alaska, or Siberia, that have never supported much human life and should be zoned for no visitation. But it's another story in places that are already inhabited. These areas are going to be developed one way or another. Americans need to understand that the world's population is increasing and that international environmental destruction is happening at an alarming rate. We can't tell local people that they can't profit from their own natural areas, and tourism represents a far nicer alternative than, say, logging or strip mining.

Q: What about the impact we have on local people?

A: We may feel guilty about visiting a place, but thousands of people are literally begging for ecotourism to come to their areas. The money that is generated goes a long way. More important is that once the people see how much we care about their place and how they can benefit from that, keeping it wild becomes more important to them as well. Every member of Audubon is an emissary for conservation.

managed in the right way. Hence tourism is sometimes suggested as a less damaging form of development by environmentalists who fear the Third World may be committing 'ecocide' through logging or other activities that use up natural resources in their struggle to survive.[9] For NGOs and governmental agencies concerned with the environment *and* development, it provides an apparent solution; or in the words of USAID a 'win–win' situation.[10] The community can earn money from tourists appreciative of the natural environment, and this money can support the community in their existing way of life. The direct benefits to the local populations concerned may include the opportunity to work in conservation, salaries paid from aid funds, revenue from ecotourism, and sometimes infrastructural benefits such as schools and medical facilities. The material benefits are evident. In a sense it is true that everyone wins – the environment is preserved, and local people benefit. The 'UN Year of Ecotourism' in 2002, marks the growing prominence of this strategy.

One lauded example of this approach is the aid-funded Communal Areas Management Programme for Indigenous Resources (Campfire) that operates in Zimbabwe. Here limited game hunting and ecotourism are organised for tourists, with the revenues contributing to supporting the rural populations in these areas. The environment – the wildlife and its environs – are preserved, and the population can benefit through the tourism revenues. Rather than being a problem, then, such tourism is put forward as *appropriate development*, as a solution, in rural Zimbabwe.

It is worth asking the question as to whether this symbiosis is *per se* a good thing in this context. The holistic approach advocated by Goodwin and central to the claims of New Moral Tourism looks at the relationship between people and nature, and makes a virtue out of not separating the two. People, it is held, are a part of nature, and their relationship to it is key to sustainability. The New Moral Tourism philosophy of community involvement is presented as benevolent to rural populations reliant on the land – they are able to gain some benefits and 'live in harmony with parks'. But why not, in so far as environments are vital (be it for tourism, the economy or science), separate people from their environment? Why not offer them something better than a life close to nature? People are part of nature, but the dominant part. Humanity has developed on the basis of harnessing nature and organising it around needs and wants. There is nothing moral or positive in encouraging specific groups of people to remain in a traditional relationship to their land, rather it seems to reflect low horizons as to what is possible with regard to development. Surely, it is one manifestation of a narrow development agenda that should be challenged rather than lauded as innovative and even moral.

Tourism and culture – symbiosis?

And just as tourism can be seen to have a symbiotic relationship with the environment, so too with culture. Local cultures can attract tourism, generate revenue and potentially make those cultures viable. These 'traditional' cultures are typically seen as 'very green' by environmentalists.[11] Advocates of New Moral Tourism engage in a self-conscious attempt to promote the inter-generational passage of culture, in terms of craft skill and religious beliefs etc. There is an unstated assumption that such a project is worthy – that encouraging cultural diversity around the world, promoting local identity, is desirable and very much part of an imperative of sustainable development.

Such cultural tourism in the Third World may be on the one hand simply buying into a desire for 'culture', or a sense of spirituality on the part of the tourist – it can be good business. However, it is considered to be much more than this. It is also held to contribute to the imperative to conserve cultures and cultural diversity.

The relationship of tourism to the environment on the one hand and culture on the other, are really two sides of the same coin. Modern environmental thinking regards traditional, rural cultures as exemplary of more sustainable practice with regard to the environment. By this reasoning, anything that supports traditional cultures is good for the environment and anything that conserves the environment is good for these cultures. Such a perspective is evident at the highest levels of global environmental governance. Principle 22 of the Rio Declaration on Environment and Development states that:

> Indigenous populations and their communities and other local communities have a vital role in environmental management and development because of their knowledge and traditional practices.[12]

Cultural tourism along these lines has been endorsed at the highest levels of global government. For World Bank President James Wolfensohn, 'culture can be justified for tourism, for industry and for employment, but it must be seen as an essential element in preserving and enhancing national pride and spirit'.[13] Schemes that key into cultural tourism markets in order to engender development often make reference to this broader perspective of cultural conservation or, in this case, 'enhancing national spirit'.

However, the relationship between tourism and culture is seen as being a fragile one. UNESCO points to the potential in, but also to the cautious attitude needed towards, utilising cultural tourism in this way:

> Cultural Tourism can encourage the revival of traditions and the restoration of sites and monuments. But unbridled tourism can have the opposite effect. Here there is a real dilemma. Is there not a risk that the boom in cultural tourism, by the sheer weight of numbers involved, may harbour the seed of its own destruction by eroding the very cultures and sites that are its stock in trade?[14]

Elsewhere the acceptance of tourism's role in cultural conservation is rather more grudging than cautious. Egyptologist Rainer Stadelman argues that, 'Tourism is already a catastrophe. But we have to admit that without tourism there would be no public interest, and without that, there would be no money for our work.'[15] What is important here is tourism's role in maintaining aspects of culture through *making culture pay*. Advocates of New Moral Tourism bemoan the demise of cultural diversity due to broader economic and social trends, and see New Tourism, be it sometimes grudgingly, as a way of turning interest in different cultures into a means of encouraging their viability. Tourism is in such circumstances portrayed as a necessary evil.

However, much of the advocacy of tourism's role in maintaining culture is more upbeat than this. One author lists potential positive impacts of cultural tourism: building community pride, enhancing a sense of identity, encouraging revival or maintenance of traditional crafts, enhancing external support for minority groups and preservation of their culture, broadening community horizons, enhancing local and external appreciation and support for cultural heritage.[16]

One important expression of the discussion of culture and tourism's ability to aid in its conservation is the UNESCO list of sites designated as World Heritage Sites. The rapid expansion of the World Heritage list is one result of a profound sense of unease with what is regarded as 'cultural levelling' – seen as the increasing dominance of a homogenous, global culture that sweeps diversity away. Mass Tourism is seen as complicit in this. These sites are predominantly 'cultural' rather than 'natural' in that they generally relate to the built environment as an expression of human cultures. Towns, villages as well as monuments, buildings etc. can be designated as World Heritage Sites.

Designation can be important, and is much sought after as it can assist in the promotion of tourism to an area – people are more likely to want to visit a World Heritage Site. Applications to become a World Heritage Site if accepted carry an obligation to preserve the sight, and use revenue gained from this status to assist in this. Hence the awarding of World Heritage Status is an example of tourism contributing to the maintenance of important aspects of cultural diversity. As one UNESCO report has it, tourism can 'help keep alive' and even 'encourage the revival of traditional cultures'.[17]

Holiday snaps – Danny Glover's Great Railway Journey

As part of a BBC series on 'Great Railway Journeys', American actor and ambassador for the UN Danny Glover travelled on a train towards Kita in Mali as part of his journey in central Africa. On the train he met two women traders. One had red eyes through crying – they had fared badly on this trip. She said it as a matter of honour that she was returning empty-handed. She recounted to Glover that she had sons who wanted to go to America, and asked whether he could help.

Glover's trip went on to visit a group of the Doggon people. On seeing the musical celebration of the hunt, Glover commented that 'they have a powerful sense of who they are, and where they come from' even if they were poor.

He watched the Doggon mask dances, unchanged over many generations except that they now attracted some tourist dollars. Glover had a Doggon mask in his American home and felt a connection with the people he was visiting.

Danny Glover's wonderment at the culture he encountered – his sensitivity and attempts to empathise with his hosts – could have meant little to the women on the train desperate for money, desperate to see their sons get to America. And the Doggon chief explained to Glover that he was fearful that education, not tourism or commercialisation, would put paid to the old traditions.

A clear example of the promotion of passing culture down, one genera-tion to the next, is the UCOTA (Ugandan Community Tourism Association) project. This project oversees the 'Heritage Trail' scheme developed by British charity Action For Conservation Through Tourism (ACT). It posits tourism as a way of conserving ancient cultures, rather than principally the environ-ment or wildlife. The Heritage Trail is organised around the four ancient kingdoms within Uganda, stressing the diversity within the country and the possibilities for attracting tourists interested in this. Here, the theme is 'making culture pay' rather than 'making nature pay'. Moreover, strengthening and even reviving of cultural identities is an explicit aim of the Heritage Trail.

In is worth noting in passing that the Heritage Trail scheme is in part a reaction to the collapse of the gorilla watching market in Uganda, an activity that had previously attracted aid funding. Gorilla watching, however, has suffered greatly since the tragic murders of tourists near the Rwandan border in 1998. The collapse of this risky market raises a question over the appli-cation of the term 'sustainable tourism' to such a scheme. Reliance on the natural world in this way is often fraught with uncertainties especially in places such as Uganda's borders, where conflicts over resources are played out. Gorilla tourism may have helped to sustain the gorilla population, but proved to be totally unsustainable for the Ugandan economy.

The rise of 'community'

The view that tourism can combine development with conservation of the environment and cultures is significant given the criticisms made of some conservation projects in the past. For example, in 1994, the World Wide Fund For Nature (WWF) used the slogan: 'He's destroying his own rain forest. To stop him, do you send in the army or an anthropologist?' in a fund-raising advertisement.[18] This approach was rightly criticised for its assumption that 'the west knows best' with regard to the value of the rain-forest, and its assumption that the solution lay with the outside expert rather than the local population. One NGO, Survival International, pointed out that this slogan implied a 'nature first' approach, with local people pushed into the background.[19]

A further example is that of the wildlife sanctuary developed in the Ngorogoro crater in Tanzania in the 1980s. Here the nomadic inhabitants of the crater, the Masai, were pressured and offered inducements to leave the crater, and they agreed to leave to live on its edge. They subsequently found it difficult to subsist here, yet were unable to return to the more fertile soil in the crater. The project responsible for this was condemned by environmentalists and others close to the Masai and knowledgeable about their way of life. The Masai were victims of the conservation-first approach, which left them without the most basic means of subsistence.

Indeed, such criticisms of environmental NGOs are not uncommon. Writing in 1991, one author goes so far as to describe US environmental institutions as tending 'to see environmental protection in isolation from the social context, . . . [They] would soon convert Costa Rica's forests into fenced-off Green museums surrounded by starving peasant families'.[20]

The conservation-oriented NGOs, then, have been accused of putting the environment before people; a 'win–lose' scenario. It is in response to this sort of criticism that many of the NGOs have reformulated and re-presented their projects using the language of 'community' and 'culture' alongside that of 'environment' and 'nature'. Conservation International, World Wide Fund For Nature and other major NGOs, previously criticised for a form of conservation that ignored local populations, are now keen to trumpet their community credentials, as are smaller NGOs and campaigns.

More precisely, they explicitly link environmental concern with the well-being of the local community. If the environment is conserved, the viability of a rural existence dependent on that environment is also improved. This is, then, the holistic approach adopted by today's NGOs. It draws together environment and culture, the former being central to the latter for poor rural communities, rather than treating the two in isolation. Nature-based tourism has a key role here in generating benefits for communities derived from conservation.

The clearest expression of the moral authority of 'community' over nature *per se* is the growth of community tourism. Ecotourism, a few years ago the vogue for tourism-for-conservation projects, now has its critics within the milieu of NGOs and campaigning organisations. It is important, they argue, for tourism, as any other form of development, to be sensitive first and foremost to *community* needs.

Today community tourism is considered by many to be the state of the art in ethical tourism. Community tourism has been most clearly set out by British campaigning group Tourism Concern, who see it as 'people first' rather than 'environment first'. Their *Community Tourism Guide* offers a brief analysis of this type of tourism and lists holidays that conform to this. They define community tourism as that which 'involves genuine community participation and benefits'.[21] *The Community Tourism Guide* goes on to argue: 'It is only by putting people at the centre of the picture that true conservation solutions will be found.' This is revealing – conservation remains the aim, but local communities have to be brought onside. But what if a community did not want to put the author's 'true conservation solutions' first? What if they prefer to leave the community in search of more lucrative jobs in hotels and in the cities? What if they view their culture as restrictive? What if they want to break away from the poverty in their community? Community tourism provides answers for conservationists confronted with the accusation that they are only concerned with the environment, but fewer to the question of development itself.

'People first' community tourism puts an emphasis on community democ-racy – involving the community in decision making. However, democracy presupposes that there are substantially different options for people to choose freely between. In the poor rural communities where community tourism is mainly advocated, these choices do not exist. Instead agendas are set by those offering aid or investment.

Perhaps we have championing of community tourism abroad precisely because of the sense that community has diminished in the developed world through urbanisation, mobility and globalisation of culture. So why not go in search of the missing elements of the modern existence? Tourists can buy community-based holidays staying in rural villages throughout the Third World. There is no reason why people should not enjoy this type of holiday, or why it might not be an appropriate development option in certain circumstances. However, community tourism, like ecotourism, is more than a holiday. It has become part of a degraded development agenda that cannot see beyond development as Third World societies living off their natural resources.

The community tourism agenda has been widely adopted by major conservation-oriented NGOs, governmental aid organisations and campaigns. Ecotourism is rhetorically no longer solely about conservation of the environment – there has to be seen to be consultation with and benefits for the local community – 'ecommunity tourism', perhaps? The International Ecotourism Society define ecotourism as 'responsible travel to natural areas that conserves the environment *and improves the well-being of local people*'.[22] USAID support projects that claim to integrate conservation and development activities in many Third World countries. These, they claim, provide alternatives to encroaching into protected areas to hunt, log and farm. Furthermore, 'a new group of stakeholders with a vested interest in protecting parks' is created. It is clearly important for them to offer benefits to host communities, as 'potential local resistance to setting aside forest and fishing areas for conservation can often be softened by employment and income-producing opportunities Ecotourism can generate'.[23] This suggests that sponsorship of ecotourism is after all to do with environmental imperatives, and that the small economic benefits to communities are instrumental to this aim – to clear the way for its acceptance within developing world communities. In similar fashion, Conservation International argue that 'All projects need to integrate the conservation of neighbouring ecosystems with the creation of economic opportunities for local residents . . .'. Furthermore, 'the development of an Eco Tourism project depends on building a local constituency that has a vested economic interest in protecting their natural resources'.[24]

Community participation is also sometimes seen as having an educative function for the host population as well as for the tourist. Erlet Cater argues that community involvement 'extends beyond economic survival, environmental awareness and sociocultural integrity to allow *appreciation by the community of their own resources*' (my italics).[25] Ecotourism, then, preaches conservation ethics to host populations, and provides material incentives to back this up. Not dissimilar to this are the Wildlife Clubs established by some NGOs in parts of rural Africa to promote a sense of the value of wildlife conservation amongst children.

Holiday snaps – 'ecommunity tourism'? Some ecotourism definitions[26]

WWF: 'tourism to protect natural areas, as a means of economic gain through natural resource preservation . . .'

The International Ecotourism Society: 'Purposeful travel to natural areas to understand the culture and natural history of the environment, taking care not to alter the integrity of the ecosystem, while producing economic opportunities to make the conservation of natural resources beneficial to local people.'

PATA (Pacific Area Tourism Association): 'The Eco Tourist practices a non-consumptive use of wildlife and natural resources and contributes to the local area through labour or financial means aimed at directly benefiting the conservation issues in general, and to the specific needs of the locals.'

International Union for the Conservation of Nature (IUCN): 'Ecotourism is environmentally responsible travel and visitation to relatively undisturbed natural areas, in order to enjoy and appreciate nature (and any accompanying cultural features – both past and present) that promotes conservation, has low negative visitor impacts, and provides for beneficial active socio-economic involvement of local populations.'

It is not surprising that the community tourism agenda, incorporating the language of culture and community, has been taken up so widely by the advocates of ecotourism. It is no more than a subtler version of the same eco-first philosophy. As Tourism Concern's *Community Tourism Guide* states:

> If conservationists want [communities] to say 'no' to harmful development, they must offer them alternative means of feeding their families. Tourism may be that alternative. In many places, tourism is a central pillar of emerging alliances between local communities and conservation organisations.[27]

But benefits of this sort come at a price, be they labelled ecotourism or community tourism. The role of poor rural communities as guardians of the environment is, in a sense, reinforced. The choice they are faced with is to accept aid or investment on the terms offered, or not at all. The aid is effectively tied to a particular conception of what these societies are capable of achieving. The relationship of poor, rural communities in the Third World to their environment remains intact, but now this relationship is viewed not as limiting their economic prospects but as a successful example of development.

Campfire

One of the most discussed, celebrated, and occasionally criticised, community tourism projects is the Communal Areas Management Programme for Indigenous Resources (Campfire) project in Zimbabwe (mentioned earlier). A brief examination of the project reveals the problems and limitations of New Moral Tourism as a development tool.[28]

Campfire was introduced in rural parts of Zimbabwe in the late 1980s. The project effectively encourages wildlife conservation, but on the basis of establishing opportunities for the local communities to benefit from it. The two stated aims of Campfire are: to conserve natural resources in the communal areas; to increase income-earning opportunities in poor communities.

The aim is to make nature pay; to establish the natural heritage as a viable basis for the local economy. Primarily 'making nature pay' is to be achieved by allowing limited hunting tourism, the bounties from which contribute towards community benefits and the control of wildlife numbers. Indigenous hunting is outlawed. Simply, the bounties paid by hunter tourists, it is argued, can generate more value for local communities than the meat from animals they hunt themselves. In effect, then, tourism generates income that enables local communities to live in harmony with wildlife and their natural environment.

A village elder has explained the scheme thus:

> It is as if we are farming wild animals, but instead of getting meat and skins for them, we get money that the tourists pay to see them. That is why we must look after our wild animals.[29]

Such schemes facilitate some economic development, but one that is based on the existing relationship of the population to the land. Benefits can be delivered, but the limits on these are set by an imperative to maintain the people's way of life defined by their relationship to the natural environment. Hence in a sense development is redefined as working around rather than in any way challenging people's direct dependence on the land.

Like the village elder quoted, we tend to define the success of such development projects by what they deliver to the local populations. In the case of Campfire there is some debate on its performance. However, it is worth also considering what such forms of aid implicitly rule out for that population. The kind of thoroughgoing development that would increase the independence of local people from their immediate environment is discounted as unrealistic or, commonly, inappropriate.

Campfire is funded through a variety of governmental and non-governmental aid agencies. It involves a complex system of administration, the shape of which is itself revealing. The Zimbabwean Department of National Parks and Wildlife Management (DNPWM) works closely with a variety of NGOs including the WWF and the Zimbabwe Trust. At a national level the Ministry of Local Government, Rural and Urban Development

(MLGRUD) are concerned with income generation and are legally responsible for the management of wildlife within limits set by the DNPWM, whose primary role is conservation. At a local level, Rural District Councils (RDCs) play a key role in implementation.

Whilst community participation is well established within the Campfire literature, some have questioned the reality. One report suggests that the aim is 'not to allow local communities to choose what to do with "their wildlife", but to teach them how to manage it in the manner DNPWM sees fit'.[30] DNPWM has the sanction of withdrawing Campfire status if their policy is not complied with. Campfire is less a case of 'co-management', as claimed, and more a case of 'persuasion'. Ultimately that power to persuade is strong, backed up by the financial authority of the NGOs. It is not fanciful to suggest that poor communities will accept the limited benefits of Campfire on the terms available, rather than questioning this at the risk of cutting off these benefits.

The process is driven from the northern NGOs, their conception of wildlife and the community's relationship to it – it is the western NGOs that provide the finance for Campfire. Whilst it is true that the Zimbabwean government ministry is centrally involved through the DNPWM, there is in fact a special Campfire unit within this department which has tended to bypass local DNPWM offices. Instead, the Zimbabwean trust, an NGO, has set up its own network of regional and local offices which liaise directly with participating RDCs.

One report describes this process thus: 'The general trend in Campfire has been to set up its own set of structures, which operate in conjunction with, but separate from, the existing central and local government structures.'[31] This process extends not only to RDCs but also to wards and villages, which are supposed to set up special ward wildlife committees, including both 'traditional' and 'modern' leaders.

A question rarely posed about this and other such projects is the extent to which they may be establishing lines of authority and finance that bypass and potentially undermine a country's existing structures of governance. For example, USAID have channelled aid directly to the local level, to the RDCs, rather than through the DNPWM at a national level. If this is the case, it would suggest that Campfire may not be contributing to *political* capacity development, but undermining national sovereign structures. Put simply, if local areas identify with structures other than the Zimbabwean government for sources for finance, this may weaken the authority and efficacy of the government.

Campfire has delivered tangible gains to participating communities. These include the building of schools and health centres. Yet there is also evidence that benefits have been very limited in scope and variable in different regions.[32] Moreover, the basis of the benefits has been to tie rural communities to the task of natural resource management – they effectively receive benefits in return for co-operating and assisting in the management of wildlife. Yet in an important sense it is their relationship to their natural environment that defines their poverty. The relationship between people and their environment is not in any sense transformed, but instead is reinforced, and subsidised, through aid funding and the hunting bounties of wealthy tourists.

It may be argued by some that rural Third World communities have very little other than their natural resource endowment upon which to base their development prospects,[33] but the point is that talking up this sort of development as 'sustainable' or 'appropriate' at best makes a virtue out of a necessity and at worst lowers development aspirations by tying them to the land. Put another way, it may be the case that regions in the Third World have a comparative advantage in 'nature', but this is precisely what defines their Third World status. Organising development around this involves a very limited conception of development.

Another way to view projects such as Campfire is as a form of tied aid. Tied aid has been widely criticised as being as much about supporting one's own economy as benefiting developing countries, as the aid is conditional on purchasing from the donor country. Clare Short, Britain's Minister at the Department for International Development, has spoken out against it. Yet here considerable sums of aid to Zimbabwe are tied, not to purchases, but to a conservation philosophy generated in the milieu of western NGOs, increasingly fêted by western governments.

Advocates of ecotourism and community tourism see such schemes as morally superior to Mass Tourism because of their ability to generate environmentally sensitive development. But this resides on a particular view of development and of the developing world. In reality New Moral Tourism is no more of a solution to less-developed status than more traditional forms of tourism. Indeed, it carries a set of assumptions that may be more limiting to the aspirations for development in the Third World. Zimbabwe has a gross national product per person of around US$ 650 compared with US$ 14,000 for the UK. Championing development on the basis of the existing relationship between people and their environment rules out the kind of transformative economic development that could make inroads into such gross inequality.

Holiday snaps – Sunungukai Camp (excerpt from *The Community Tourism Guide*)[34]

'The first Campfire tourism project to be fully run by local communities. Sunungukai, in northern Zimbabwe on the edge of the Umfurudzi Safari Area, consists of four traditional style chalets and offers rugged hiking, fishing, birdwatching and wildlife plus the chance to see Bushman paintings, meet traditional healers, experience village life and buy local crafts. The camp is run by a locally elected committee. $10 per night.'

Solomon Islands

The Solomon Islands provide a good example of the 'tourism for conservation' approach. The development charity Oxfam, under their 'Community Aid Abroad' scheme, offer tourists the chance to 'celebrate and preserve the unique biodiversity of the rainforests' and 'enjoy the warm hospitality of remote village communities and share their mountain lifestyle which has evolved in harmony with the rainforest ecosystem'. Ecotourism is seen as an important source of income in this area as it supports 'their desire to preserve, not destroy the forest'.[35]

Of course, the chance to visit the Solomon Islands is one many would relish. However, it seems to be tied here to discouraging the inhabitants of the island from acting in a manner deemed damaging to the natural environment. Aid given on this basis embodies a conception of the Solomon Islands as a bastion of nature, and this view may have implication for how wider developments are viewed; developments that may be deemed destructive of nature.

Central Ghana Project

Another fêted tourism-for-conservation project is Conservation International's Central Ghana Project. The project received the sought-after British Airways Tourism For Tomorrow award in 1999. Whilst other forms of industrial development, it is argued, may damage the environment, 'Ecotourism continues to be a positive alternative to destroying Ghana's rainforest. Ecotourism can help create jobs and business opportunities for local communities while building appreciation for a country's natural heritage and culture.'[36] The project includes an interpretative exhibit, 'Hidden Connections', highlighting the biological and cultural connections between the rainforest and people. Yet it is worth asking what this 'natural heritage and culture' actually represents. These 'biological and cultural connections' are presented almost as a natural order of things. It is an order that eschews change and involves abject poverty for the majority.

It would seem that a particular conception of culture is presented, one that is only holistic in the sense that it ties people to the limits of the land and rules out any change in the relationship between people and their land. The aid creates some benefits for some people, but at the expense of defining a large swathe of the country as a 'conservation area'. The limited horizons implicit in this version of culture are all too evident in a country lacking basic industries; a country being forced to live off its natural capital.

Belize

Belize is well known as a haven for nature, and has a high profile as an ecotourism destination. This is not surprising given that well over 40 per cent of its land mass has been given over to reserves and conservation areas, established by wealthy environmental NGOs. These NGOs wield immense power over the shape of the Belize economy – ecotourism figures prominently, as it is regarded as less damaging than the ways the people of Belize made a living prior to the reserves. One author describes the environmentalists active in Belize as the 'New Missionaries', exuding a moral authority to preserve nature, with the communities on Belize something of an afterthought.[37] The shaping of Belize by environmental NGOs is evident in the relocation of Maya residents to make way for the Cockscomb jaguar sanctuary, with the hiring of a local schoolteacher as sanctuary director as part of the deal.

The US-based Audubon Society, by agreement, established the Belize Audubon Society, which administers eight protected areas established under the National Parks Systems Act. NGOs and environmental groups have played a key role in shaping the Belize economy in an image acceptable to their own ecocentric view.

Having established ecotourism as a key industry in Belize, environmentalists have been able to manipulate eco-sentiment to pressure the Belize government to stop developments unfavourable to the environmental agenda. Some groups publicised the logging of Belize's Columbia Forest Reserve in the US and organised a campaign of letters, faxes and e-mails from 'would-be' tourists to Belize, saying they would be going elsewhere until the logging stopped.[38]

Pro-poor tourism

In the UK, the Department for International Development (DfID) also sponsors a scheme to promote small-scale, sustainable tourism in the Third World. The literature sees tourism as a means of addressing the 'pro-poor' agenda to establish basic needs for the world's poorest people. One of the justifications given for utilising tourism in this way is that, 'Tourism products can be built on natural resources and culture, which are often the only significant assets the poor have.' This lack of resources defines their underdeveloped status. The pro-poor tourism approach works around this to engender a limited development in rural areas. Such a development can hardly be held up and celebrated; it seems to reflect rather low horizons within the development debate over what is possible.[39]

Indeed, one department workshop paper begins with a quote from the World Wide Fund For Nature suggesting that nature constrains the extent to which poverty can be tackled: 'Sustainable Tourism is tourism and associated infrastructures that, both now and in the future, operate within *natural* capacities for the regeneration and future productivity of natural resources'[40] (my italics). Yet in what sense are there natural limits in the fashion implied? The limits to development in the Third World are better regarded as *social* in nature rather than rooted in natural processes. They are a product of unequal economic and political relationships, and more immediately, the burden of debt and the dearth of inward investment and of aid itself. Few in the more developed countries live within limits defined by their specific relationship to their immediate environment in the way the Department for International Development's advisors seem to be advocating here in the name of sustainability.

The pro-poor initiative is also discussed in terms of new benefits for the poor. Of these, 'Economic benefits are only one (very important) component – social, environmental and cultural costs and benefits also need to be taken into account.'[41] One of these cultural impacts is that tourism may employ more women than men, thus creating greater quality between the sexes in the areas concerned. Whilst in the abstract greater equality must be a good thing, the attempt to engineer equality through aid in this way must itself raise some ethical questions.

'Pro-poor tourism' takes issue with what it rightly sees as 'a defensive or protectionist approach: "preserving local culture", "minimising costs"' in the language of sustainable tourism. Pro-poor tourism seeks to expand opportunities. In addition, it sees itself as a broader approach than community-based tourism, because it also prioritises the links between the poor community and the formal sector. As such pro-poor tourism has much to be said for it as an approach to tackling rural poverty. However, it too reflects low horizons on the development front. Rather than constituting an attempt to liberate people from the constraints of their environment, pro-poor tourism organises around these constraints. Worthy, perhaps, but hardly an inspiring vision of what is possible.

There is nothing wrong with ecotourism as a commercially viable way of making money. However, as with the DfID scheme, it has come to appear as the form of tourism associated with innovative development, and has been taken up by a variety of NGOs with this justification. When ecotourism is presented as a dynamic development strategy, it is important to point out that it offers relatively little in the way of benefits. Moreover the outlook behind ecotourism, that views human development only in terms of the pre-existing relationship between man and nature, precludes a discussion of broader developmental needs. That ecotourism can be celebrated and presented in an upbeat way as a means to development suggests that some have lowered their sights with regard to development possibilities. Real development requires transformation of the relationship to the natural world. Ecotourism development projects take the relationship as it exists and institutionalises it through lines of funding from the developed world.

The projects mentioned are examples of a broader trend that involves New Moral Tourism as a tool in development, or to be more precise, integrated conservation and development. The impulse to encourage Third World states to adopt conservation is far from new. However, the discourse and the practice are very different today from the past.

What's new?

In the past colonial attitudes to conservation tended to see the west as superior, and as such with the authority to impose its own solutions without recourse to consultation. These solutions included the imposition of national parks and game reserves, in line with the colonial conception of, for example, Africa as a haven for wilderness and wildlife (alongside its role as a source of cheap raw materials and its strategic importance). The reserves served to meet the recreational need of the colonisers. Many of the same 'natural' areas remain a priority for tourism-for-conservation projects today.

Post independence, national parks and reserves were generally not a priority for newly liberated African states that sought to shed Third World status through wide-scale economic development. Some western environmentalists tried to encourage the Third World to see 'the virtue of living off the income of their natural resources, not the capital'.[42] In many ways, the growth of tourism-for-conservation strategies represents the victory of environmental concern over the development ambitions of many of Africa's post-colonial regimes.

One author puts this down to the growing importance of the tourism industry and the growing ability of environmental NGOs such as Conservation International, WWF, Nature Conservancy, Smithsonian Institute and government agencies such as USAID to shift priorities through lines of funding.[43] One could add to this the failure of national development projects in Africa. It is only in this context that 'living off natural resources' can be presented to Africans and others in the Third World in any kind of positive manner.

In contrast to the colonial period, today advocates of New Moral Tourism present themselves as defenders of cultural diversity in poor rural areas of the Third World. Globalisation must be resisted and diverse cultures must survive the modern assault, it is held. The advocates of New Moral Tourism see themselves as radicals, at the cutting edge of development, and are strongly critical of colonial attitudes to the Third World. Yet their defence of culture and the environment carries assumptions that help perpetuate the inequality established in the colonial era. Indeed, in the name of defending cultural diversity, equality barely gets a look in. The language of diversity has become a bastion against change. Development, appropriate development, is limited to that which is possible *given* existing culture; culture expressing the relationship between people and their environment.

Sustainable development: sustainable for whom?

The nature-based tourism adopted by NGOs is generally discussed as being sustainable, both in its own terms and with reference to Mass Tourism, as it is based on the conservation of natural resources rather than their transformation. It is worth referring to the most commonplace definition of sustainable development – that originating in the Brundtland Commission, and popularised at the UN Rio Earth Summit in 1992, to consider the implications of this. Here, sustainable development is defined as, 'development that meets the needs of the present without compromising the ability of future generations to meet their own needs'.

In the USA GNP per capita is $29,240 whilst in Kenya it is $964 (allowing for estimated differences in purchasing power).[44] Are the needs of either country's population met? And who is to decide what these 'needs' are? The UN? Or perhaps the World Bank, who service the 'needs' of developed economies by removing greater wealth from Africa through debt repayments than is injected in through meagre aid budgets? This reality – that some people meet their needs by preventing others from meeting theirs – is overlooked.

In fact, there is an emphasis in sustainable development circles on meeting '*basic* needs'. Basic needs – the absolute minimum that people need to survive – are obviously a priority. Ecotourism can provide revenue to help provide for these basic needs. However, it also seems to preclude going far beyond basic needs, as to do so would contradict the environmental conservation agenda that is at the heart of Campfire and other projects seeking to utilise tourism as a development tool. Needs over and above basic ones may conflict with the priority given to the environment. They would involve transformative change on the scale that is ruled out as damaging to the environmental and diluting of cultures.

New Moral Tourism's economic claims

The arguments in favour of New Moral Tourism as a development tool are not restricted to its claimed benign characteristics for culture and the environment. Alongside this it is argued to possess positive characteristics for economic development relative to more mainstream multinational-led hotel and resort development.

Community tourism, ecotourism and related new tourisms present smallness of scale as central to their advantages. Small, more personal and, importantly, locally owned businesses are seen as more beneficial to the local population. Indeed, smallness of scale has always been an important plank of environmental thinking, epitomised by Schumacher's *Small is Beautiful* written in the early 1970s. In this context it is important to consider the benefits of small-scale tourism.

Advocates argue that it can deliver greater gains to local populations, as it is based within communities and is therefore more likely to draw upon locally owned businesses in accommodation, food, crafts etc. The argument goes that $10 spent within an eco- or community tour will go further than $10 spent on a traditional package.

Formally there is, of course, a case for this. Undoubtedly large, foreign-owned hotels have shareholders to worry about, and have no obligation to reinvest in the areas in which they are based. They typically repatriate much of the profit made and often prefer standardisation to local sourcing. Wages may be low for locally recruited staff, whilst managerial staff from the hotel company's base country may be more handsomely paid. On the other hand, an ecotourist purchasing locally produced crafts directly from the people who made them may be fairly confident that their money goes to local people.

Yet large hotels enable large numbers of people to visit, therefore whilst a dollar for dollar comparison may flatter community tourism, the potential to generate foreign exchange is much greater for more mainstream hotels. Community tourism and ecotourism, by their very nature, must involve small numbers of people (usually in rural areas) and therefore the overall impact is limited. Also, when the community tourism dollar does accrue to local populations, this is more likely to be in the informal economy, and in craft production. The extent to which this can contribute to development on any wide-scale level has itself to be questioned. Indeed, in extreme ecotourism, based in the wilderness, there may be nothing to spend money on anyway.

Polly Pattullo rightly argues in *Last Resorts: the Cost of Tourism in the Caribbean* that an important limitation of tourism's development potential is its failure to link up with local suppliers. She cites examples in the Caribbean where Irish potatoes and Florida orange juice take precedence over domestically grown yam, breadfruit, mangoes and bananas. Various initiatives have tried to create more linkages within Caribbean economies, such as the 1992 'Time For Action' initiative of the West Indian Commission. This document urged that 'agriculture, manufacturing and tourism be developed on a symbiotic basis'.[45]

Pattullo identifies a vital point – that if tourism is to have the greatest beneficial effect, then other related industries need to develop. Yet the large-scale transfers of capital needed to bring this about have never materialised. Also, the development of a broader range of industries that tourism could complement in a more diversified economy would require the very things advocates of New Moral Tourism eschew – modern industry utilising modern technology, and on a large scale.

In place of this, New Moral Tourism cites greater self-sufficiency through linkages to the local economy in terms of crafts and 'sustainable' agricultural produce. The latter usually refer to production processes that are economically inefficient and labour intensive. Stephen Page points out that 'the more self-sufficient an economy, the greater the revenue retained', and hence the greater the multiplier – the knock-on impact of tourist spending.[46] This is true, formally, but ultimately trade is beneficial. It is the basis on which trade takes place that is problematic. Self-sufficiency is not a virtue in and of itself – economies isolated from the world economy are amongst the poorest.

Critics of multinational investment-led tourism development, typically in resorts and hotels, cite a low multiplier effect, lack of linkages with local industries, low wages and a reliance on foreign capital as evidence of its limitations. Yet in a sense these things comprise the meaning of 'less developed' or 'Third World' – the inability to trade on any sort of equal basis, lack of local capital and a capitalist class, lack of auxiliary and complementary industries. What does the new approach represent in relation to this imperfect state of affairs? In many ways it is more of a retreat than an advance. Tourism development in the past may have carried many limiting features, but New Moral Tourism turns away from development itself.

Mass Tourism is no panacea, but it is an industry, like any other, that can and does improve the lot of people where it is located. It can bring substantial benefits to developing economies. One (rare) author prepared to explicitly defend Mass Tourism cites the Dominican Republic, that has in the last twenty-five years built a tourism industry based on all-inclusives, as an example.[47] All-inclusives are, for New Moral Tourists, the worst, least ethical forms of tourism as they are large scale and focus on the comforts of the guests rather than the needs of the environment or community directly. But because of their large scale and orientation towards the market they attract many tourists – some 2.6 million in 1999, generating US$ 2.5 billion in revenue and 140,000 direct jobs. This represents around 15 per cent of GDP and 30 per cent of export earnings in the Dominican Republic.

No one would argue that tourism in the Dominican Republic is about to transform this poor country. However, the country is a lot better off with than without its Mass Tourism industry. Moreover ecotourism can provide nowhere near the same levels of foreign exchange earnings or job creation.

One might also cite The Gambia as a country that has benefited from Mass Tourism. The Gambia receives around 85,000 international tourists annually, with tourism accounting for about 11 per cent of GDP and some 7,000 jobs directly and indirectly. In the absence of regular scheduled services, the regular charter holiday flights have created an important transport link with knock-on effects on the economy.[48]

Yet opposing the hotel-based all-inclusive holiday developments commonplace in The Gambia has become a *cause célèbre* for western advocates of New Moral Tourism, concerned over the corrosive impact on culture as well as the limited economic effect. Gambia remains amongst the poorest countries in Africa, but it is likely it would be poorer still if tourism development had been subject to the restrictions desired by the New Moral Tourism lobby.

Mass Tourism has played a significant role in generating development in countries such as Spain. There, broader industrial growth and the integration of Spain into the world economy in the 1960s and 1970s set the context in which revenue from tourism contributed to the transformation of the Spanish economy. Such positive contextual factors are not evident in the Dominican Republic today. This suggests that it is the broader context of the Dominican Republic that limits the effect of tourism development, rather than the type of tourism itself. Broader economic and political factors determine the possibilities, or lack of them, for tourism to contribute to economic transformation.

It is difficult to accept that small-scale tourism has anything significant to offer the development agenda. In attempting to improve on the development performance of the traditional sector, community tourism simply replaces old problems with new ones – high leakages are replaced with low foreign exchange potential. More importantly, in reacting to the injustices evident in the way large multinational businesses operate, the advocates of New Moral Tourism turn away from substantial developments towards those that create few 'negative impacts'. The problem is that they succeed in producing few positive ones either.

Impacts are generally interpreted in a negative fashion anyway by the New Moral Tourism – environmental change is interpreted as environmental destruction and cultural change is seen as being in a negative direction. Moreover, the negative consequences accruing to some people in the course of tourism development are customarily elevated above the positive consequences of tourism-led development for many others. One report reminds us of 'the many dangers that lurk behind the cash that travellers inject into the world economy', and that it 'wreak[s] havoc with local cultures and environments'.[49] Development based upon tourism undoubtedly brings uncertainties and insecurities, leaves some as winners and others as losers and creates problems with regard to the environment – although havoc is surely too strong a word. However, alongside the 'havoc' comes new opportunities. Development is both a process of destruction and creation simultaneously. New Tourism advocates focus on destruction, and fails to recognise the benefits accruing from development.

Who benefits?

Finally, it is worth stressing that this conservation agenda that purports to offer development through nature-based tourism is rarely a matter of free choice for Third World countries. North–South relations are characterised by an intense inequality in political and economic life. Third World debt and the fortress approach to travellers from the Third World coming to the west are facets of this. Many of the same institutions championing tourism as a means of achieving 'win–win' – USAID, the United Nations, the World Bank – are the same ones enforcing debt repayments, presiding over structural adjustment programmes and denying access to richer countries to Third World peoples.

Structural adjustment programmes, organised through the World Bank, IMF and United Nations, are a key feature of this inequality. Structural adjustment effectively means that some developing world countries can benefit from better terms on their debt, but that in turn the country must follow a particular set of economic policies. One author argues that the creditor organisations 'virtually control the economies' of many developing countries.[50] These policies in general involve an opening up to international investment and competition. In this way the rich can pull rank on the poorer nations through managing the process of debt repayment and debt write-offs.

Of course, this may be viewed as an eminently sensible option. The conditions put on the assistance may be seen as ensuring better governance and greater economic growth. However, it amounts to an infringement of sovereignty. Moreover the debt situation is a consequence of the subservient position of the South in the first place – it is an expression of the subjugation of the poorer nations by the rich. To 'offer' to write off debts in this way is less an act of generosity and more one of imposing one's will using debt as leverage.

Some of the very institutions presiding over this state of affairs are keen to promote ecotourism and conservation generally as a benefit for Third World environments and societies. The hypocrisy is stunning. The use of New Moral Tourism to promote a more green form of development is advocated by the UN . . . yet aid budgets of UN members remain miserly.

A number of prominent conservation schemes operate through debt for nature swaps (see Table 2).[51] These operate on principles similar to the structural adjustment programmes. The swaps operate by large international NGOs purchasing debt that Third World countries owe to commercial banks. The banks will sell at a discounted rate, as they may consider the chances of repayment to be uncertain. The Third World government then effectively owes the NGO instead of the bank. The NGOs renegotiate the debt on more favourable terms for the country, normally over a longer period and the interest on the bonds now held by the NGO is used to support conservation programmes. Conservation areas established are typically sites for the development of ecotourism of some kind, as a means of making these areas more economically viable. On maturity of these bonds, the principal can act as an endowment for the local NGO operation.

Table 2 Examples of debt for nature swaps

Date	Country	Cost $	Purchaser	Face value of debt $	Conservation funds $	Total external debt $
1989	Madagascar	950,000	WWF	2.1m	2.1m	3.9b
1990	Madagascar	2.5m	CI	5m	5m	
1990	Madagascar	446,000	WWF	919,000	919,000	
1989	Zambia	454,000	WWF	2.3m	2.3m	7.2b
1991	Ghana	250,000	CI/SI/DDC	1m	1m	10.5b

Source: WWF and World Bank Reports

CI = Conservation International
DDC = Debt For Development Coalition
SI = Smithsonian Institute
WWF = World Wildlife Fund
M = millions/b = billions (US)

For the conservation NGOs this mechanism enables them to promote conservation in the countries concerned. This is in part a response to the conservation imperative generally, but also there is a concern that pressure to service debt repayments has pushed Third World countries into environmentally unsustainable practices such as logging, clearing forests for cash crops and cattle grazing, part of a drive to produce exportable goods to help the debt situation.

Debt for nature swaps may be seen in the context of the 'win–win' scenario set out earlier – they deliver environmental conservation and some debt is written off at the same time. However, the debt reduction is conditional on a limited transference of sovereignty to the NGO concerned. Also, it is tied to an acceptance of turning over swathes of territory to national parks and other conservation areas. These areas can generate further finance through ecotourism. Ecotourism revenue can provide the incentive for local communities to support these schemes – it is how, as set out earlier, the NGOs can win friends and influence people. Ultimately, though, the assistance is given on condition that it contributes to reinforcing a pre-existing relationship to the land in rural areas – conservation is the aim, substantial development is eschewed.

The debt for nature swaps show how the broader inequalities between North and South enable NGOs to establish a conservation agenda in the Third World. The idea that by tagging ecotourism on to conservation in some way makes this a 'development' agenda too is to diminish the idea of development still further.

Innocent fun on Redcar Beach, Cleveland, in the UK. (Photo: Joanna Williams)

Postscript
Re-presenting tourism

Tourism is being recast as an arena for moral proscription and critical self-awareness. I have argued that this version of tourism is not a progressive development, as its advocates claim. Rather, it is a recipe for wariness and personal guilt in an arena traditionally associated with innocence, fun and a footloose and fancy free attitude.

The root of the modern critique of tourism is to see the relationship between people and nature as brittle. Tourism has been a prime target for environmental concern, because for the tourist the environment is often part of what they enjoy, often in large numbers. The tremendous growth of opportunities to travel and enjoy the environment – the beach, warm climates, snow-covered mountains – is regarded by the critics as a threat.

The critique has also moved on to increasingly present the relationship between different peoples, host and tourist, as adversarial. Each is defined by their different cultures, with the emphasis on *different*. There has been a boom in advice for the tourist, presupposing that they require guidelines and codes to negotiate this world of difference. Accompanying this, a dim view is taken of those who ignore the advice and treat a holiday as just a holiday.

The result of this is that tourism is changing, and not for the better. In different times package tourism was more likely to have been regarded simply as a welcome respite from the rigours and proscriptions of everyday life. Today, wariness and caution better describe the sentiment encouraged by the New Moral Tourism. Our holidays have become a vessel into which we are encouraged to pour environmental angst and fears of globalisation. New Moral Tourists travel with a sense of personal mission, as tourism is recast as philanthropy towards the hosts and a 'unique experience' for the tourist.

The ire directed against tourism is misplaced. The growth of Mass Tourism has been a mark of real progress in modern society. Many can travel abroad for leisure when only a couple of generations ago foreign travel was a rarity for most people. New opportunities have opened up as the holiday companies have expanded to ever more destinations. This has not been at the expense of those hosting the growing numbers of tourists. Tourism to the resorts on the Spanish Costas – for tourism's critics surely near the top of the 'unethical' list – have contributed to economic growth that has enabled the Spanish themselves to travel increasingly as tourists. Indeed, Torremolinos can no longer be viewed as a colony of British lager louts with little thought for locals – a large proportion of tourists late on in the season are the Spanish themselves.

In poorer countries the gulf between tourist and host is more marked. The response to this from advocates of New Moral Tourism is to call for more 'sensitive' tourism – tourism that tries to avoid displays of wealth such as comfortable hotels and cameras. But whilst tourism may bring together people with different access to wealth and opportunity, it did not create this inequality. Get rid of the all-inclusive resorts, the conspicuous displays of wealth and privilege, and poor people are if anything worse off.

It is true that tourism can bring together people of very different backgrounds – the ultra-rich and the poor, the latter perhaps cleaning rooms for the former in some grand hotel. But for the most part, tourists are ordinary people seeking out good weather, good food, fun and perhaps a little taste of a different way of life. It is probably something they have worked and saved hard for. The portrayal of tourists, still today, as rather thoughtless people who contribute to the exploitation of the places they visit, is belittling. It also trivialises any discussion of poverty and how to tackle it. In fact, the solutions offered up by the New Moral Tourism lobby are less concerned with material wealth and more with cultural difference. More often than not a preoccupation with the latter gets in the way of addressing the dearth of the former (or even seeing it as a dearth at all).

Campaigners for New Moral Tourism like to champion the cause of those who have lost out in some way from resort developments. Yet they are reticent to balance this against the many benefits derived from tourism developments. Their agenda is preservationist not only with regard to the natural world, but also with regard to culture. Individuals and their aspirations for a better life are distorted by the preoccupation with an iconic *culture*, cast in china, in need of protection from other cultures, other people.

Whilst the moralisation of tourism sees only the differences between people, we should regard travel, for leisure, for education or for business, as part of a common culture. This is not in the sense that we all engage in it, but that there can be few who do not have the *aspiration* to travel for leisure. Aspiration is important – it provides a link between culture as *what is* and culture as *what could be*. But culture in the latter sense is often obscured by a debate that assumes the differences between peoples as their defining characteristic, even in the prosaic realm of leisure travel.

And what of the backpackers – the tourists who see themselves as travellers? As we have seen, they have in many ways joined the ranks of the package tourists in the eyes of the critics. They, too, are held responsible for commercialising cultures around the world and damaging environments.

Joseph Conrad described the origins of his thirst for travel thus, in *The Heart of Darkness*:

Now when I was a little chap I had a passion for maps. I would look for hours at South America, or Africa, or Australia, and lose myself in the glories of exploration. At that time there were many blank spaces on the earth, and when I saw one that looked particularly inviting on a map (but they all look like that) I would put my finger on it and say, 'When I grow up I will go there.'

The spirit of this passage has inspired many, and although there are no more 'blank spaces' on the earth, the desire to travel remains strong. Pete Smith, a 28-year-old manager for the specialist youth travel company STA Travel, sees hundreds of customers every month buying round-the-world tickets and booking short breaks to every continent. He recalls his father telling him about his one trip abroad, a three-day coach journey to Barcelona. The Sombrero he brought back still adorns the family living room wall. For Pete himself, New York had a mythical quality in his teenage years – a city synonymous with music and the movies – the 'city that never sleeps'. Yet now short breaks across the Atlantic are commonplace for his customers, and he is planning a trip to the Big Apple. Such is the growth of international travel.[1] My three-year-old son has a Children's Map of Europe from Stanfords Map shop in London on his wall, and has already decided that Lapland, at the very edge of the map (and with a picture of Santa and his reindeer), is one place he would like to go!

The moral baggage associated with travel now threatens to shackle a spirit of adventure for travellers young and old. As travel has become a focus for moral codes, something has been lost along the way. If travel is to really be a 'life-expanding activity', or a 'unique experience' of any kind, then it has to rely on the individual, be they reckless or sensitive, impulsive or well prepared. Attempts to formalise codes of conduct, and the constant appeals for deference to the interests of the host's 'environment' and 'culture' only contribute to a spirit of caution rather than one of adventure and discovery. Today's young travellers are counselled to be New Moral Tourists – travel is to be wary, cautious and condemnatory of those who favour the resorts.

Tourism looks set to continue to grow. Notably, parts of Asia are increasingly getting the travel bug, and the economic growth to realise the ambition to travel. Even in the Third World, where people's aspirations for travel are confounded by poverty and immigration laws, numbers able to travel as tourists are increasing, be it slowly. How we interpret this growth is an open question. Is tourism to be an angst-ridden pursuit, a necessary evil, something we do but wish others wouldn't (hell, when it comes to holidays, is always other people), or is it to be guilt-free enjoyment?

There is no need to make a grand case for package tourism, rooted in some supposed positive effect on tolerance, worldliness or as the World Travel and Tourism Council bizarrely claim in its effect on promoting world peace. John F. Kennedy once stated that, 'Travel has become one of the great forces for peace and understanding in our times.' Yet to misquote a popular analysis, there have been more people killed in wars since the advent of Mass Tourism than in any other period in history, so Kennedy was wide of the mark. In fact tourism has no great moral claims to be made for it. It is not the aim here to counter the doom merchants with a celebration of leisure travel as the high point of human culture, or a boon to cultural tolerance (although the aspiration and growing ability to travel are worthy of celebration).

Tourism need only be about enjoyment, and requires no other justification. As for the moralising about tourist behaviour, how people choose to enjoy themselves is a matter for them. Some may have a preference for solitude above conviviality, the wilderness above the resort, but New Moral Tourism is disingenuous in dressing this up as having a moral basis.

Of course, footloose international travel remains, globally, the prerogative of the few, not the many. But given the chance, few would choose to stay at home. And whether abroad is Bombay or even Benidorm, that aspiration is something to be salvaged from the angst-ridden commentaries on tourism, and celebrated.

Notes

Introduction

1 J. Croall, *Preserve or Destroy: Tourism and the Environment*, London, Calouste Gulbenkian Foundation, 1995

1 Mass Tourism and the New Moral Tourist

1 *The Collins English Dictionary*, Glasgow, Collins, 1986
2 *Pocket Oxford Dictionary*, Oxford, Clarendon, 1984
3 *Shorter Oxford English Dictionary* Oxford University Press, 1973
4 R. Bray and V. Raitz, *Flight to the Sun: the Story of the Package Holiday Revolution*, London, Continuum, 2001, p. 20
5 World Tourism Organisation figures, cited in *Industry and Environment*, United Nations Environment Programme, Vol. 24, No. 3–4, July–Dec., 2001, p. 5
6 A. Poon, *Tourism, Technology and Competitive Strategy*, Wallingford, CABI, 1993
7 Ibid.
8 A. Poon, 'Competitive Strategies for a "New" Tourism', in C. Cooper (ed.), *Progress in Tourism, Recreation and Hospitality Management*, vol. 1, London, Belhaven, 1989, p. 84
9 A. Poon, *Tourism, Technology and Competitive Strategy*, Wallingford, CABI, 1993, p. 3
10 WTTC, WTO and Earth Council, *Agenda 21 for the Travel and Tourism Industry: Towards Environmentally Sustainable Development*, London, World Travel and Tourism Council, 1995
11 Ibid.
12 Tourism Concern/DfID, *Looking Beyond the Brochure*, 1999 (educational resource – video)
13 Quoted in J. Croall, *Preserve or Destroy: Tourism and the Environment*, London, Calouste Gulbenkian Foundation, 1995, p. 1
14 K. Srisang, 'Third World Tourism: the New Colonialism', *In Focus* (bulletin of campaigning NGO Tourism Concern), No. 4, 1992, pp. 2–3
15 Ibid.
16 Pope John Paul, cited in L. Purves, 'Tourists Should Not Travel Light on Morals', *The Times*, 10 July 2001
17 Proyecto Ambientale Tenerife publicity information, 1997
18 Studienkreis für Tourismus und Entwicklung (Students for Tourism and Responsibility) aims and objectives, listed on their web site at: http://www. studienkreis.org/engl/wer/ziele.html accessed on 17/05/2002
19 See The International Ecotourism Society's 'Ecotourism Explorer' advice initiative, http://www.ecotourism.org/travelchoice/investigate.html accessed on 16/05/2002

20 L. Purves, 'Tourists Should Not Travel Light on Morals', *The Times*, 10 July 2001
21 J. Griffiths, 'Tourism is Bad for our Health', *Guardian*, 8 Feb. 2001
22 G. Monbiot, 'An Unfair Exchange', *Guardian*, 15 May 1999
23 Tourism Concern Press Release, issued 14 July 2001
24 Cited in WTO, *The British Ecotourism Market, special report no. 11*, World Tourism Organisation, Madrid, 2001, p. 33
25 A. Rice, quoted in M. Wells, 'Travel Shows Portray Paradise and Hide Reality', *Guardian*, 28 August 2001
26 I. Ousby, *The Englishman's England: Taste, Travel and the Rise of Tourism*, Cambridge, Cambridge University Press, 1990, p. 130
27 Brochures for Explore, Dragoman and Encounter Overland, undated
28 Quoted in R. Coward, 'Ten Ways to Give Your Guilt a Holiday', *Guardian*, 5 Feb. 2001
29 Figures quoted in L. Barton, *Guardian, Education* supplement, 4 Sept. 2001
30 Tony Higgins, Chief Executive, UCAS (Universities and Colleges Admissions Service) quoted in World Expeditions Challenge Limited publicity, 2000
31 Trekforce brochure, undated
32 World Tourism Organisation, *Global Tourism Forecasts*, WTO, Madrid, 1995
33 World Tourism Organisation figures, cited in *Industry and Environment*, United Nations Environment Programme Vol. 24, No. 3–4, July–Dec. 2001, p. 5
34 P. Mason and M. Mowforth, *Codes of Conduct in Tourism*, Occasional Papers, University of Plymouth, 1995
35 M. Mowforth and I. Munt, *Tourism and Sustainability: New Tourism in the Third World*, London, Routledge, 1998, ch. 6
36 S. House and K. Wood, *The Good Tourist*, London, Mandarin, 1991, preface
37 G. Neale, *The Green Travel Guide*, 1st edn, London, Earthscan, 1998
38 G. Neale, *The Green Travel Guide*, 2nd edn, London, Earthscan, 1999, Introduction
39 L. Turner, and J. Ash, *The Golden Hordes: International Tourism and the Pleasure Periphery*, London, Constable, 1975
40 D. Pearce, *Tourist Development*, Harlow, Longman, 1989, p. 100
41 A. Poon, *Tourism, Technology and Competitive Strategy*, Wallingford, CABI, 1993, p. 4
42 G. Monbiot, quoted in G. Neale, *The Green Travel Guide*, 1st edn, London, Earthscan, 1998, p. xxiii
43 S. Wearing and J. Neil, *Ecotourism: Impacts, Potentials and Possibilities*, London, Butterworth-Heinemann, 1999, p. 129
44 S. Plogg, 'Why Destination Areas Rise and Fall in Popularity', in E. Kelly (ed.), *Domestic and International Tourism*, Wellesley, MA, Institute of Certified Travel Agents, 1997
45 Cited in R. Sharpley, *Tourism, Tourists and Society*, Elm, Huntingdon, 1994, p. 84
46 S. Wearing and J. Neil, *Ecotourism: Impacts, Potentials and Possibilities*, Butterworth-Heinemann, London, 1999, p. 70
47 H. Ceballos-Lascurain, *Tourism, Ecotourism and Protected Areas*, Gland, Switzerland, IUCN, 1996
48 Cited in I. Munt, 'Ecotourism or Egotourism?', *Race and Class*, Vol. 36, No. 1, July–Dec., 1994, p. 58
49 D. Birkett, 'How Good Causes Have Hijacked Holidays', *Daily Express*, 13 July 2000, p. 14
50 P.S. Valentine, 'Ecotourism and Nature Conservation: a Definition with some Recent Developments in Micronesia', in *Ecotourism Incorporating the Global Classroom, International Conference Papers*, 1992, pp. 4–9

51 W.R. Eadington, and V.L. Smith, 'Introduction: the Emergence of Alternative Forms of Tourism', in V.L. Smith and W.R. Eadington (eds), *Tourism Alternatives: Potentials and Problems in the Development of Tourism*, Philadelphia, University of Pennsylvania Press, 1992, p. 6

52 D. Brooks, *Bobos in Paradise: the New Upper Class and How They Got There*, New York, Simon and Schuster, 2000, pp. 205–206

53 H. Blumer, 'The Mass, the Public and Public Opinion', in A.M. Lee (ed.), *New Outlines of the Principles of Sociology*, New York, Barnes and Noble, 1939

54 R. Williams, *Culture and Society*, Harmondsworth, Penguin, 1961, p. 289

55 C. Cooper *et al.*, *Tourism: Principles and Practice*, 1st edn, London, Pitman, 1993, p. 103

56 L. Lencek and G. Bosker, *The Beach: the History of Paradise on Earth*, London, Pimlico, 1998

57 F. Barret, 'On the Algarve's Road to Ruin', *Independent*, 22 July 1989, p. 45

58 J. Urry, *The Tourist Gaze: Leisure and Travel in Contemporary Society*, London, Sage, 1990, p. 87

59 G. Therborn, *European Modernity and Beyond: the Trajectory of European Societies 1945–2000*, London, Sage, 1996

60 P. Corrigan, *The Sociology of Consumption*, London, Sage, 1997, p. 145

61 The dramatic analogy, in which experiences of other societies can be 'backstage', referring to an authentic experience, or 'frontstage', referring to a something acted out for a particular audience, is commonly invoked in the academic study of the sociology of tourism. It originates with the sociologist Irving Goffman. See I. Goffman, *The Presentation of Self in Everyday Life*, Harmondsworth, Penguin, 1959

62 P. Corrigan, *The Sociology of Consumption*, London, Sage, 1997, p. 145

63 See R. Sharpley, 'Tourism and Sustainable Development: Exploring the Theoretical Divide', *Journal of Sustainable Tourism*, Vol. 8, No. 1, 2000, pp. 1–19

64 Federation of Nature and National Parks in Europe, Grafenau, Germany *Loving Them to Death?*

65 See for a useful discussion R. Sharpley, 'Tourism and Sustainable Development: Exploring the Theoretical Divide', *Journal of Sustainable Tourism*, Vol. 8, No. 1, 2000, pp. 1–19

66 Ibid. and see M. Stabler (ed.), *Sustainable Tourism: From Principles to Practice*, Wallingford, CABI, 1997

67 S. Wahab and J. Pigram, *Tourism Development and Growth*, London, Routledge, 1997, ch. 3

68 Mowforth and Munt usefully term the NGOs concerned with sustainable tourism as 'socio-environmental' organisations, as opposed to simply environmental or ecological organisations. M. Mowforth and I. Munt, *Tourism and Sustainability: New Tourism in the Third World*, London, Routledge, 1998, ch. 6

69 See S. Page and R. Dowling, *Ecotourism*, London, Prentice Hall, 2002 for useful discussion of the evolution and definition of the term 'ecotourism'

70 See *Journal of Sustainable Tourism*, Vol. 8, No. 4, 2000, pp. 341–351

71 H. Ceballos-Lascurain, *Tourism, Ecotourism and Protected Areas*, Gland, Switzerland, IUCN, 1996

72 S. Page and R. Dowling, *Ecotourism*, London, Prentice Hall, 2002

73 Cited in C.G. Bottrill and D.G. Pearce, 'Ecotourism, Towards a Key Elements Approach to Operationalising the Concept', *Journal of Sustainable Tourism*, Vol. 3, No. 1, 1995, pp. 45–54

74 T.G. Acott, H. La Trobe and S. Howard, 'An Evaluation of Deep Ecotourism and Shallow Ecotourism', *Journal of Sustainable Tourism*, Vol. 6, No. 3, 1998, pp. 238–253

75 Ibid. p. 249

76 Ibid. p. 250

2 What's new?

1 M. Graham, cited in J. Croall, *Preserve or Destroy: Tourism and the Environment*, London, Calouste Gulbenkian Foundation, 1995, p. 56
2 L. Withey, *Grand Tours and Cooks Tours: a History of Leisure Travel 1750–1915*, London, Aurum, 1997, p. 144
3 Cited in D. Boorstin, *The Image: a Guide to Pseudo Events in America*, New York, Vintage, 1992 (first published 1964), p. 83
4 Ibid. p. 82
5 I. Ousby, *The Englishman's England: Taste, Travel and the Rise of Tourism*, Cambridge, Cambridge University Press, 1990, p. 187
6 J. Ruskin, cited in D. Boorstin, *The Image: a Guide to Pseudo Events in America*, New York, Vintage, 1992, p. 87
7 J. Ruskin, cited in M. Feifer, *Going Places: the Ways of the Tourist from Imperial Rome to the Present Day*, London, Macmillan, 1985, p. 167
8 Ibid. ch. 5
9 J. Myerscough, 'The Recent History of the Use of Leisure Time', in I. Appleton (ed.), *Leisure, Research and Policy*, Edinburgh, Scottish Academic Press, 1974, p. 14
10 J. Buzzard, *The Beaten Track: European Tourism, Literature and the Ways to Culture, 1880–1918*, Oxford, Clarendon, 1993
11 Rev. F. Kilvert, *Diary*, 1870, quoted in J. Croall, *Preserve or Destroy: Tourism and the Environment*, London, Calouste Gulbenkian Foundation, 1995, p. 72
12 From an article in *Blackwoods* magazine, February 1965, quoted in D. Boorstin, *The Image: a Guide to Pseudo Events in America*, New York, Vintage, 1992, p. 88
13 I. Ousby, *The Englishman's England: Taste, Travel and the Rise of Tourism*, Cambridge, Cambridge University Press, 1990, p. 89
14 Cited in D. Boorstin, *The Image: a Guide to Pseudo Events in America*, New York, Vintage, 1992, p. 88
15 Ibid.
16 Cited in P. Brendon, *Thomas Cook: 100 Years of Popular Tourism*, London, Secker and Warburg, 1991, p. 58
17 I. Ousby, *The Englishman's England: Taste, Travel and the Rise of Tourism*, Cambridge, Cambridge University Press, 1990, p. 89
18 R. Bray and V. Raitz, *Flight to the Sun: the Story of the Package Holiday Revolution*, London, Continuum, 2000
19 Lonely Planet, *Andalucia*, 2nd edn, London, Lonely Planet Publications, 2000, p. 268
20 D. Facaros and M. Paul, *Spain*, London, Cadogan, 1999, p. 603
21 L. Withey, *Grand Tours and Cooks Tours: a History of Leisure Travel 1750–1915*, London, Aurum, 1997, p. 145
22 J. Urry (1990), *The Tourist Gaze: Leisure and Travel in Contemporary Society*, London, Sage, 1992, p. 34
23 Ibid. ch. 2
24 M. Mowforth and I. Munt, *Tourism and Sustainability: New Tourism in the Third World*, London, Routledge, 1998, ch. 5
25 M. Feifer, *Going Places: the Ways of the Tourist from Imperial Rome to the Present Day*, London, Macmillan, 1985, p. 179
26 H. Marcuse, *One-Dimensional Man*, London, Routledge, 1999 (first published in 1964)
27 D. Boorstin, *The Image: a Guide to Pseudo Events in America*, New York, Vintage, 1992 (first published 1964), p. 85
28 Cited in S. Hall, 'Backpackers Hit the Tourist Trail', *Guardian*, 29 June 1999
29 L. Purves, 'Tourists Should not Travel Light on Morals', *The Times*, 10 July 2001

30 Cited in M. McGrath, 'On The Road Again', *Guardian*, 10 June 2000
31 Peter Smith, STA Travel, pers. com. Nov. 2001
32 T. Wheeler, 'Power to the Backpackers', in the *Independent, Travel* section, 7 July 2001, p. 1
33 B. Wheeler, 'Alternative Tourism – a Deceptive Ploy', in C.P. Cooper and A. Lockwood (eds), *Progress in Tourism, Recreation and Hospitality Management*, London, Belhaven, 1992
34 M. Pryer, 'The Traveller as Destination Pioneer', *Progress in Tourism and Hospitality Research*, vol. 3, 1997, p. 235
35 Ibid, p. 236
36 See Y. Apostolopoulos *et al.* (eds), *The Sociology of Tourism: Theoretical and Empirical Investigations*, London, Routledge, 1996, p. 40
37 R. Koshar, *German Travel Cultures*, Oxford, Berg, 2000, p. 2
38 Ibid. p. 2
39 D. Boorstin, *The Image: a Guide to Pseudo Events in America*, New York, Vintage, 1992
40 D. MacCannell, *The Tourist: a New Theory of the Leisure Class*, London, Macmillan, 1976
41 A. Stalbow, 'There's No Escape', *Independent on Sunday, Review* section, 7 Feb. 1999
42 Cited in S. Hall, 'Backpackers Hit the Tourist Trail', *Guardian*, 29 June 1999
43 G. Paulsen, 'My Holidays: Gary Paulsen', *Sunday Times, Travel* section, 21 July 1997
44 P. Fussell, *Abroad: British Literary Travelling Between the Wars*, Oxford, Oxford University Press, 1980
45 B. Wheeler, 'Eco or Ego Tourism: New Wave Tourism – a Short Critique', *Insights*, English Tourist Board, May, 1992, pp. 41–42
46 A. Pleumaron, excerpt from 'Eco-tourism or Eco-terrorism?', a briefing paper to the German Association for Political Economy, April 1995, at, http://www.untamedpath.com/Ecotourism/ecoterrorism.html, accessed on 22/08/2002
47 E. Cater, 'Profits From Paradise', *Geographical*, London, March, 1992
48 M. Honey, *Ecotourism and Sustainable Development: Who Owns Paradise?*, Washington DC, Island Press, 1999, p. 90
49 P.A.A. Berle, 'Two Faces of Ecotourism', *Audubon*, No. 92, 1990, p. 6
50 Cited in Tourism Concern resource pack, from Tourism Concern magazine *In Focus*, photocopy, undated.
51 M. Turco, 'The Greed Behind the Green in Tourism', *Mail and Guardian* (African English-language publication), 2001 http://www.mg.co.za/mg/africa/dec-tourtraps2.html accessed 26/09/01

3 The host

1 J. Croall, *Preserve or Destroy: Tourism and the Environment*, London, Calouste Gulbenkian Foundation, 1995, p. 1
2 Ibid.
3 Quoted in R. Cowe, 'Tourism Concedes Restraint May be Needed to Help Environment', *Guardian*, 6 Sept. 1995, p. 19
4 S. Wearing and J. Neil, *Ecotourism: Impacts, Potentials and Possibilities*, London, Butterworth-Heineman, 1999, p. 48
5 R. Prosser, 'Societal Change and the Growth in Alternative Tourism', chapter in E. Cater and G. Lowman (eds), *Ecotourism: a Sustainable Option?*, Chichester, Wiley, 1994, p. 33
6 D. Harrison and M. Price, 'Fragile Environments, Fragile Communities? An Introduction', chapter in M. Price (ed.), *People and Tourism in Fragile Environments*, Chichester, Wiley, 1996, p. 1

7 Ibid, p. 1
8 Ibid, p. 2
9 P. MacNaughten and J. Urry, *Contested Natures*, London, Sage, 1998, ch. 2, p. 19
10 D. Harrison and M. Price, 'Fragile Environments, Fragile Communities? An Introduction', chapter in M. Price (ed.), *People and Tourism in Fragile Environments*, Chichester, Wiley, 1996, p. 23
11 E. Cater, 'Profits from Paradise', *Geographical*, London, March, 1992, pp. 16–21, p. 19
12 Agenda 21, cited in L. Pera and D. McLaren, 'Globalization, Tourism and Indigenous Peoples: What You Should Know About the World's Largest "Industry"', Rethinking Tourism Project, Nov. 1999, http://www. planeta.com/ecotravel/resources/rtp/globalization.html
13 D. Harrison and M. Price, 'Fragile Environments, Fragile Communities? An Introduction', chapter in M. Price (ed.), *People and Tourism in Fragile Environments*, Chichester, Wiley, 1996, p. 9
14 C.P. Gurung and M. De Coursey, 'The Annapurna Conservation Area Project: a Pioneering Example of Sustainable Tourism', in E. Cater and G. Lowman (eds.), *Ecotourism: a Sustainable Option?*, Chichester, Wiley, 1994, p. 179
15 Ibid. p. 178
16 Pers. com. Nadia Theuma of the University of Malta, March, 2002
17 See F. Furedi, *Population and Development: a Critical Introduction*, Cambridge, Polity, 1997 for an exposition of this debate in the area of population studies
18 M. Feifer, *Going Places: the Ways of the Tourist from Imperial Rome to the Present Day*, London, Macmillan, 1985
19 See http://www.ecotourism.org/travelchoice/sustain.html accessed on 16/05/2002

4 Tourists

1 J. Croall, *Preserve or Destroy: Tourism and the Environment*, London, Calouste Gulbenkian Foundation, 1995, p. 56
2 H. Muller, 'The Thorny Path To Sustainable Tourism Development', *Journal of Sustainable Tourism*, Vol. 2, No. 3, 1994, p. 134
3 Guideline 2, 'Tourism With Insight' code of conduct for tourists (undated)
4 J. Jones, 'Can Tourism Be Green', *Green World*, March, 1996
5 J. Krippendorf, *The Holiday Makers: Understanding the Impacts of Leisure and Travel*, Oxford, Heinemann, 1987, p. xiv
6 G. Neale, *The Green Travel Guide*, 1st edn, London, Earthscan, 1998
7 P. Ghazi and J. Jones, *Downshifting: a Guide to Happier, Simpler Living*, London, Coronet, 1997
8 J. Jones, 'Can Tourism Be Green', *Green World*, March, 1996
9 G. Neale, *The Green Travel Guide*, 1st edn London, Earthscan, 1998, Foreword
10 Tourism Concern leaflet, undated
11 Centre for Environmentally Responsible Tourism publicity leaflet, undated
12 Friends of Conservation 'Conservation Code', undated
13 Survival International publicity leaflet, undated
14 ASTA's *Ten Commandments of Eco-Tourism*, http://www.astanet.com/travel/ecotravel.asp?r=/travel/ecotravel.asp, accessed on 11.10.02
15 ACAP, *Minimum Impact Code, Annapurna Conservation Area Project*, Kathmandu, King Mahendra Trust, WWF, 1989
16 J. Krippendorf, *The Holidaymakers: Understanding the Impact of Leisure Travel*, Oxford, Heinemann, 1987, ch. 16
17 D. McLaren, *Rethinking Tourism and Ecotravel: the Paving of Paradise and What You Can Do To Stop It*, West Hartford, CT, Kumarian Press, 1998

18 G. Monbiot quoted in G. Neale, *The Green Travel Guide*, 1st edn, London, Earthscan, 1998, p. xxiii
19 Cited in M. McGrath, 'On The Road Again', in *Guardian, Weekend* supplement, 10 June 2000
20 F. Furedi, *The Culture of Fear: Risk Taking and the Morality of Low Expectations*, London, Cassell, 1997
21 S. House and K. Wood, *The Good Tourist*, London, Mandarin, 1991, p. 77
22 M. Mann, *The Community Tourism Guide: Exciting Holidays for Responsible Travellers*, London, Earthscan, 2000, p. xii
23 Ibid. p. 4

5 The cultural sensibilities of the New Moral Tourist

1 Sociologist Erik Cohen has explored the idea of tourists centring themselves, spiritually, in different locations. Explorers centre themselves with the Other, whilst package travellers' spiritual centre is their home. See E. Cohen, 'A Phenomenology of Tourist Experiences', *Sociology*, No. 13, 1979, pp. 179–201
2 The Imaginative Traveller (undated), General Information, UK edition, p. i
3 Ibid.
4 J. Urry, *The Tourist Gaze: Leisure and Travel in Contemporary Society*, London, Sage, 1990, p. 95
5 S. Guild, 'My Journey With the Bedouin', *Independent, Travel* supplement, 23 Aug. 1999
6 M. Brown, *The Spiritual Tourist*, London, Bloomsbury, 1998, p. 2
7 See N. Graburn, 'Tourism: the Sacred Journey', in V. Smith (ed.), *Hosts and Guests: the Anthropology of Tourism*, 2nd edn, Philadelphia, University of Philadelphia Press; D. MacCannell, *The Tourist: a New Theory of the Leisure Class*, 2nd edn, New York, Shocken Books, 1989
8 D. Brooks, *Bobos in Paradise: the New Upper Class and How they Got There*, New York, Simon and Schuster, 2000, pp. 206–207
9 See M. Feifer, *Going Places: the Ways of the Tourist from Imperial Rome to the Present Day*, London, Macmillan, 1985; J. Urry, *The Tourist Gaze: Leisure and Travel in Contemporary Society*, London, Sage, 1990
10 J. Urry, *The Tourist Gaze: Leisure and Travel in Contemporary Society*, London, Sage, 1990, p. 3
11 G. Therborn, *European Modernity and Beyond: the Trajectory of European Societies 1945–2000*, London, Sage, 1996
12 *Flâneur* is the name given to dandy Parisian street strollers of the nineteenth century, who exhibited their style on Paris's broad boulevards
13 See J. Urry, *The Tourist Gaze: Leisure and Travel in Contemporary Society*, London, Sage, 1990, p. 26
14 D. MacCannell, *Empty Meeting Grounds: the Tourist Papers*, London, Routledge, 1992
15 G. Dann, 'Tourism Motivation: an Appraisal', *Annals of Tourism Research*, 2, 1981
16 J. Krippendorf, *The Holiday Makers: Understanding the Impacts of Leisure and Travel*, Oxford, Heinemann, 1987, p. xiv
17 C. Rojek, *Decentring Leisure: Rethinking Leisure Theory*, London, Sage, 1995, ch. 3
18 C. Ryan, *Recreational Tourism: a Social Science Perspective*, London, Routledge, 1991, p. 148
19 C. Anderson, *Our Man In Goa*, BBCTV, 1995
20 S. Woolaston, *Guardian*, 24 Feb. 1996, p. 50
21 E. Pearce, 'Goa for a song, Not a Sign of Despair', *Guardian*, 15 Feb. 1995, p. 20

22 Pers. com. with Gregoire Le Divillec of the Worldwrite charity, involved in youth exchange programmes and development issues, 16 February 1997

23 D. McClaren, *Rethinking Tourism and Ecotravel: the Paving of Paradise and What You Can Do to Stop It*, West Hartford, CT, Kumarian Press, 1998

24 L.P. Hartley, *The Go Between*, London, Heinemann, 1963

25 E.H. Carr, *What is History?*, London, Penguin, 1989

26 R. Hewison, *The Heritage Industry*, London, Methuen, 1987

27 Proyecto Ambiental, promotional literature, 1997

28 R. Hewison, *The Heritage Industry*, London, Methuen, 1987, p. 8

29 R. Williams, *Keywords*, London, Fontana, 1988, p. 147

30 See R. Hewison, *The Heritage Industry*, London, Methuen, 1987 for an analysis of this trend in the area of heritage and tourism, and see F. Furedi, *Mythical Past, Elusive Future: History and Society in an Anxious Age*, London, Pluto, 1992 for a more general analysis

31 R. Williams, *Keywords*, London, Fontana, London, 1988, p. 87

32 R. Williams, cited in Robert C. Young, *Colonial Desire: Hybridity in Theory, Culture and Race*, London, Routledge, 1995, p. 44

33 E. Cohen, 'Towards a Sociology of International Tourism', *Social Research*, No. 39, 1972, p. 165

34 I. Silver, 'Marketing Authenticity in Third World Countries', *Annals of Tourism Research*, 20, 1993, p. 10

35 D. Lodge, *Paradise News*, London, Penguin, 1991

36 D. Nash and V. Smith, 'Anthropology and Tourism', *Annals of Tourism Research*, 18, 1991, pp. 12–25, p. 13

37 A. Poon, *Tourism, Technology and Competitive Strategy*, Wallingford, CABI, 1993

38 D. MacCannell, *The Tourist: a New Theory of the Leisure Class*, London, Macmillan, 1976, ch. 5

39 I. Goffman, *The Presentation of Self in Everyday Life*, Harmondsworth, Penguin, 1959

40 Tanzanian Cultural Tourism Programme brochure, undated, written, designed and edited by Stephen H. Fisher.

41 D. Nash and V. Smith, 'Anthropology and Tourism', *Annals of Tourism Research*, 18, 1991, p. 13

42 D. Nash, *Anthropology of Tourism*, Oxford, Pergamon, 1996, p. 11

43 Ibid. p. 26

44 T. Nunez, 'Touristic Studies in Anthropological Perspective', in V. Smith (ed.), *Hosts and Guests: the Anthropology of Tourism*, Philadelphia, University of Pennsylvania Press, 1989, p. 266

45 D. MacCannell, *Empty Meeting Grounds: the Tourist Papers*, London, Routledge, 1992, pp. 3–4

46 For example, L. Turner and J. Ash, *The Golden Hordes: International Tourism and the Pleasure Periphery*, London, Constable, 1975, p. 129; D. Nash, *Anthropology of Tourism*, Oxford, Pergamon, 1996

47 C. Ryan, *Recreational Tourism, A Social Science Perspective*, London, Routledge, 1991, p. 148

48 D. Nash, *Anthropology of Tourism*, Oxford, Pergamon, 1996, p. 27

49 G. Richards, *Developing and Marketing Crafts Tourism*, Tilburg, ATLAS, 1998

50 D. MacCannell, *Empty Meeting Grounds: the Tourist Papers*, London, Routledge, 1992, p. 19

51 See for example C. Tickell, foreword in E. Cater and G. Lowman (eds), *Ecotourism: a Sustainable Option?*, Chichester, Wiley, 1994

52 Ibid.

53 K. Malik, *The Meaning of Race*, London, Macmillan, 1996, ch. 6

54 J.S. Mill on Bentham and Coleridge, cited in R. Williams, *Culture and Society*, Harmondsworth, Penguin, 1961, p.74

55 D. Nash, *Anthropology of Tourism*, Oxford, Pergamon, 1996, p. 126
56 C. Lévi-Strauss, *Triste Tropiques*, Harmondsworth, Penguin, 1955, p. 385
57 C. Lévi-Strauss, *The Naked Man*, translated by J. and D. Weightman, London, Harper Row, 1981 (originally published in 1971), p. 636
58 D. MacCannell, *Empty Meeting Grounds: the Tourist Papers*, London, Routledge, 1992, p. 18
59 C. Lévi-Strauss, *The View From Afar*, translated by Joachim Neugroschel and Pheobe Hoss, Harmondsworth, Penguin, 1987 (originally published in 1983), p. 24
60 C. Lévi-Strauss, *Structural Anthropology 2*, Harmondsworth, Penguin, translated, 1978 (originally published 1973), p. 362
61 Ibid. p. 360
62 D. MacCannell, *Empty Meeting Grounds: the Tourist Papers*, London, Routledge, 1992, p. 295
63 D. Nash, *Anthropology of Tourism*, Oxford, Pergamon, 1996, p. 35
64 A.F. Jurdao, cited in D. Nash, *Anthropology of Tourism,* Oxford, Pergamon, 1996, pp. 34–35
65 G. Wheatcroft, cited in J. Urry, 'The Tourist Gaze and the Environment', *Theory, Culture and Society*, Vol. 19, 1990, p. 24
66 J. Jones, 'Can Tourism be Green', *Green World*, March, 1996
67 G. Gmelch, 'Crossing Cultures: Student Travel and Personal Development', *Journal of Inter Cultural Relations*, 1997, Vol. 21, No. 4, pp. 475–490
68 D. Nash, *Anthropology of Tourism*, Oxford, Pergamon, 1996

6 Travelling for a change

1 T. McNaught, 'Mass Tourism and the Dilemmas of Modernisation in Pacific Island Communities', *Annals of Tourism Research*, 9, 1982, p. 360
2 M. Mann, *The Community Tourism Guide: Exciting Holidays for Responsible Travellers*, London, Earthscan/Tourism Concern, 2000, p. 3
3 K. Robins, 'Global Times: What in the World's Going On?' in P. du Gay (ed.), *Production of Culture/Cultures of Production*, London, Sage/Open University, 1997
4 P. MacMicheal, *Development and Social Change*, London, Sage, 1996, p. 11
5 V. Kinnaird, U. Kothari and D. Hall, 'Tourism: Gender Perspectives', chapter in V. Kinnaird and D. Hall (eds), *Tourism: A Gender Analysis*, Chichester, Wiley, 1994, p. 13
6 D. McLaren and L. Pera, 'Globalization, Tourism and Indigenous Peoples: What You Should Know About the World's Largest "Industry"', the Rethinking Tourism Project, Nov. 1999, http://www.planeta.com/ecotravel/resources/rtp/globalization.html accessed on 23/05/02
7 Ibid.
8 J. Krippendorf, *The Holiday Makers: Understanding the Impacts of Leisure and Travel*, Oxford, Heinemann, 1987, p. 56
9 L. Turner and J. Ash, *The Golden Hordes: International Tourism and the Leisure Periphery*, London, Constable, 1975
10 Survival International, 'Tourism and Tribal Peoples', publicity leaflet, 1995
11 D. Nash, 'Tourism as a Form of Imperialism', in V. Smith (ed.), *Hosts and Guests: the Anthropology of Tourism*, Philadelphia, University of Philadelphia Press, 1989
12 Ibid, p. 38
13 Ibid. pp. 5–6
14 E. Cohen, 'Towards a Sociology of International Tourism', *Social Research*, 39, 1972
15 V. Kinnaird, U. Kothari and D. Hall, 'Tourism: Gender Perspectives', chapter in V. Kinnaird and D. Hall (eds), *Tourism: a Gender Analysis*, Chichester, Wiley, 1994, p. 13

16 See for example P. MacMicheal, *Development and Social Change*, London, Sage, 1996

17 P. Pattullo, *Last Resorts: the Cost of Tourism in the Caribbean*, London, Cassell, 1996, p. 41

18 A. Poon, *Tourism, Technology and Competitive Strategy*, Wallingford, CABI, 1993

19 M. Mowforth and I. Munt, *Tourism and Sustainability: New Tourism in the Third World*, London, Routledge, 1998; J. Butcher, 'Sustainable Development or Development?', chapter in M. Stabler (ed.), *Tourism and Sustainability: Principles to Practice*, Oxon, CABI, 1997

20 P. MacNaughten and J. Urry, *Contested Natures*, London, Sage, 1998, p. 60

21 Ibid., pp. 80–84

22 D. Pepper, *Modern Environmentalism*, London, Routledge, 1996, p. 85

23 Evidence of ethical consumerism in tourism can be found in P. Wight, 'Environmentally Responsible Marketing of Tourism', chapter in E. Cater and G. Lowman (eds), *Ecotourism: a Sustainable Option?*, Chichester, Wiley, 1994

24 Z. Bauman, *Intimations of Postmodernity*, London, Routledge, 1992, p. 49

25 N. Hertz, *The Silent Takeover: Global Capitalism, and the Death of Democracy*, London, William Heinemann, 2001

26 See C. Leadbetter, 'Power to the Person', *Marxism Today*, 14–19 Oct. 1998; S. Hall, 'Brave New World', *Marxism Today*, 24–29 Oct. 1998

27 F. Fukuyama, *The End of History and the Last Man*, London, Penguin, 1992

28 F. Mort, 'The Writing on the Wall', *New Statesman and Society*, 12 May 1989, pp. 40–41 cited in C. Lury, *Consumer Culture*, London, Polity, 1996, p. 254

29 Ibid.

30 A. Giddens, *Beyond Left and Right: the Future of Radical Politics*, London, Polity Press, 1994, p. 5

31 See N. Lewis and J. Malone's introduction to V.I. Lenin's *Imperialism: the Highest Stage of Capitalism*, London, Pluto, 1996

32 IUCN, *Caring For The Earth*, Gland, Switzerland International Union for the Conservation of Nature, 1991

33 WWF (1995), 'Environmentally Adapted Tourism and Ecotourism', cited in L. Aronsson, *The Development of Sustainable Tourism*, London, Continuum, 2000, p. 47

34 A. Poon, *Tourism, Technology and Competitive Strategy*, Wallingford, CABI, 1993, p. 15

35 A. Giddens, *Modernity and Self Identity: Self and Society in the Late Modern Age*, Cambridge, Polity, 1991

36 For an exposition for the former view see M. Featherstone, *Consumer Culture and Postmodernism*, London, Sage, 1991, and for the latter see J. Heartfield, *Need and Desire in the Post Material Economy*, Sheffield, Sheffield Hallam University Press, 2000

37 D. MacCannell, *Empty Meeting Grounds: the Tourist Papers*, London, Routledge, 1992, p. 3

38 Ibid. pp. 21–22

7 New Moral Tourism, the Third World and development

1 G. Neale, *The Green Travel Guide*, London, Earthscan, 1998

2 US Agency for International Development, *Win-Win Approaches to Development and the Environment: Ecotourism and Biodiversity Conservation*, Centre for Development Information and Evaluation, USAID, Washington DC, July 1996

3 Conservation International Press Release, 'CI Wins Tourism For Tomorrow Award', 4 Feb. 1999, at http://www.conservation.org/WEB/NEWS/PRESS REL/99–022413HTM date 23/4/99

4 K. Lindberg and D.E. Hawkins (eds), *Ecotourism for Planners and Managers*, North Bennington, VT, Ecotourism Society, 1993

5 Table cited in UNEP, *Industry and Environment*, Vol. 24, No. 3–4, July–Dec., 2001, p. 15

6 H. Goodwin, 'Tourism and Natural Heritage, a Symbiotic Relationship?', chapter in M. Robinson, P. Long, N. Evans, R. Sharpley and J. Swarbrooke (eds), *Environmental Management and Pathways to Sustainable Development*, Sunderland, Business Education Publishers Limited, 2000. Also G. Budowski, 'Tourism and Conservation: Conflict, Coexistence or Symbiosis', *Environmental Conservation*, No. 3, 1976

7 Audubon Society, 'The Ethics of Ecotravel', http://magazine.audobon.org/features0009/ethics.html accessed on 16/05/2002

8 Ibid.

9 E. Cater, 'Ecotourism in the Third World – Problems and Prospects for Sustainability', chapter in E. Cater and G. Lowman (eds), *Ecotourism: a Sustainable Option?*, Chichester, Wiley, 1994, p. 84

10 US Agency for International Development, *Win-Win Approaches to Development and the Environment: Ecotourism and Biodiversity Conservation*, Center for Development Information and Evaluation, USAid, Washington DC, July 1996

11 J. Jones, 'Can Tourism Be Green', *Green World*, March, 1996

12 United Nations, *Report to the United Conference on Environment and Development*, Rio de Janiero, 3–14 June (Vol. 11), United Nations, New York, 1992

13 Quoted in A. Molstad *et al.*, *Sustainable Tourism and Cultural Heritage: a Review of Development Assistance and its Potential to Promote Sustainability*, report for Norwegian World Heritage Organisation, 1999, p. 1

14 Ibid. p. 1

15 Ibid.

16 Ibid. p. 7

17 A. Molstad, K. Lindberg, D. Hawkins & W. Jamieson, *Sustainable Tourism and Cultural Heritage: A Review of Development Assistance and its Potential to Promote Sustainability*, report for Norwegian World Heritage Organisation, 1999, p. 2

18 Cited in M. Mowforth and I. Munt, *Tourism and Sustainability: New Tourism in the Third World*, London, Routledge, 1998, p. 17

19 Ibid.

20 J. Carriere, 'The Crisis in Costa Rica: an Ecological Perspective', in D. Goodman and M. Redclift, *Environment and Development in Latin America: the Politics of Sustainability*, Manchester, Manchester University Press, 1991, p. 198

21 M. Mann, *The Community Tourism Guide*, London, Earthscan, 2000, p. 19

22 Ibid, p. 27

23 US Agency for International Development, *Win-Win Approaches to Development and the Environment: Ecotourism and Biodiversity Conservation*, Center for Development Information and Evaluation, USAID, Washington DC, July 1996

24 Conservation International Field Reports, *Ecotourism*, http://www.conservation.org/WEB/FIELDACT/C-C_PROG/ECON/ECOTOURHTM accessed on 23/11/99

25 E. Cater, 'Ecotourism in the Third World – Problems and Prospects for Sustainability', chapter in E. Cater and G. Lowman (eds), *Ecotourism: a Sustainable Option?*, Chichester, Wiley, 1994, p. 85

26 Quoted in C. Goeldner *et al.*, *Tourism: Principles, Practices and Philosophies*, New York, Wiley, 2000, p. 556

27 M. Mann, *The Community Tourism Guide: Exciting Holidays for Responsible Travellers*, London, Earthscan, 2000, p. 27

28 This section, on the Campfire project, is based on information from: Plan Afric,

Appendix to Report for Community Action project, report for Ministry of Public Service, Labour and Social Welfare, Zimbabwe, 1997

29 The oldest lineage head of the Purros valley families talking to anthropologist M. Jacobsohn in P. Bonner, *Living World*, Winter, 1993

30 Plan Afric, *Appendix to Report for Community Action project, report for Ministry of Public Service, Labour and Social Welfare*, Zimbabwe, 1997

31 Ibid.

32 Ibid.

33 E. Cater, Introduction, in E. Cater and G. Lowman (eds), *Ecotourism: a Sustainable Option?*, Chichester, Wiley, 1994

34 M. Mann, *The Community Tourism Guide*, Earthscan, London, 2000, p. 168

35 Oxfam Community Aid Abroad Tours, http://www.caa.org.au/travel/tours/solomons.html

36 Press release: CI Wins Tourism for Tomorrow Award, Feb. 24, 1999, accessed at *http://www.conservation.org/WEB/NEWS/PRESSREL/99-0224B.HTM*, on 23/11/99

37 A. Sutherland, *The Making of Belize: Globalisation in the Margins*, Bergin and Garvey, Westport, 1998, ch. 9, 'The New Missionaries'

38 Ibid.

38 Department for International Development, Business Partnerships Unit, Tourism Challenge Fund leaflet, 1999

40 Cited in 'Sustainable Tourism and Poverty Elimination', a report on the workshop of 13 Oct. 1998 held by the Department for the Environment, Transport and Regions and the Department for International Development, in preparation for the UN Commission on Sustainable Development, http://www.igc.org/csdngo/tourism/tour_ukgov.htm accessed on 23/05/2002

41 Department for International Development, Business Partnerships Unit, Tourism Challenge Fund leaflet, 1999

42 J. McCormick, *The Global Environmental Movement*, Chichester, Wiley, 1995, p. 49

43 D. Omotayo Brown, 'Debt-funded Environmental Swaps in Africa: Vehicles for Tourism Development?', *Journal of Sustainable Tourism*, Vol. 6, No. 1, 1998, pp. 69–79

44 World Bank figures for 1998, cited at http://www.worldbank.org/depweb/english/modules/economic/gnp/databig.htm accessed on 25/5/2002

45 P. Pattullo, *Last Resorts: the Cost of Tourism in the Caribbean*, London, Cassell, 1996, p. 39

46 S. Page and R. Dowling, *Ecotourism*, London, Prentice Hall, 2002, p. 155

47 R. Sharpley, 'In Defence of (Mass) Tourism', chapter in M. Robinson *et al.* (eds), *Reflections on International Tourism: Environmental Management and Pathways to Sustainable Tourism*, Sunderland, Business Education Publishers Limited, 2000

48 Ibid.

49 M. Turco, *Mail and Guardian* (English language African newspaper), http://www.mg.co.za/mg/africa/dec-tourtraps2.html accessed on 26/09/01

50 N. Hildyard, 'The Big Brother Bank', *Geographical*, 26–28 June 1994; also see R. Potter *et al.*, *Geographies of Development*, London, Prentice Hall, 1999, ch. 7, especially pages 166–169, for a good account

51 D. Omotayo-Brown, 'Debt-funded Environmental Swaps in Africa: Vehicles for Tourism Development?', *Journal of Sustainable Tourism*, Vol. 6, No. 1, pp. 69–79, 1998

Postscript

1 Pete Smith, STA Travel, London, pers. com., Nov. 2001

Select bibliography

Acott, T., H. La Trobe and S. Howard, 'An Evaluation of Deep Ecotourism and Shallow Ecotourism', *Journal of Sustainable Tourism*, Vol. 6, No. 3, 1998, pp. 238–253

Akama, J.S., 'Western Environmental Values and Nature-Based Tourism in Kenya', *Tourism Management*, Vol. 17, No. 8, 1996, pp. 567–574

Apostolopoulos, Y., S. Leivadi and A. Yiannakis (eds), *The Sociology of Tourism: Theoretical and Empirical Investigations*, London, Routledge, 1996

Aronsson, L., *The Development of Sustainable Tourism*, London, Continuum, 2000

Bauman, Z., *Intimations of Postmodernity*, London, Routledge, 1992

Blamey, R.K., 'Ecotourism: the Search for an Operational Definition', *Journal of Sustainable Tourism*, Vol. 5, No. 2, 1997, pp. 108–130

Bleasdale, S., 'Charity Challenges and Adventure Tourism in Developing Countries – Who Dares Wins', in M. Robinson, P. Long, N. Evans, R. Sharpley and J. Swarbrooke (eds), *Motivations, Behaviour and Tourist Types: Reflections on International Tourism*, Sunderland, Business Education Publishers Ltd, 2000

Blumer, H., 'The Mass, the Public and Public Opinion', in A.M. Lee (ed.), *New Outlines of the Principles of Sociology*, New York, Barnes and Noble, 1939

Boorstin, D., *The Image: a Guide to Pseudo Events in America*, New York, Vintage, 1992 (first published 1964)

Bray, R., and V. Raitz, *Flight to the Sun: the Story of the Package Holiday Revolution*, London, Continuum, 2001

Brooks, D., *Bobos in Paradise: the New Upper Class and How They Got There*, New York, Simon and Schuster, 2000

Brown, M., *The Spiritual Tourist*, London, Bloomsbury, 1998

Budowski, G., 'Tourism and Conservation: Conflict, Coexistence or Symbiosis', *Environmental Conservation*, No. 3, 1976, pp. 27–31

Burns, P., *An Introduction to Tourism and Anthropology*, London, Routledge, 1999

Burns, P., and A. Holden, *Tourism: A New Perspective*, London, Prentice Hall, 1995

Buzzard, J., *The Beaten Track: European Tourism, Literature and the Ways to Culture, 1880–1918*, Oxford, Clarendon, 1993

Carr, E.H., *What is History?*, London, Penguin, 1989

Cater, E., 'Profits From Paradise', *Geographical*, London, March, 1992, pp. 16–21

Cater, E., and G. Lowman (eds), *Ecotourism: a Sustainable Option?*, Chichester, Wiley, 1994

Ceballos-Lascurain, H., *Tourism, Ecotourism and Protected Areas*, Gland, Switzerland, IUCN, 1996

Cohen, E., 'Towards a Sociology of International Tourism', Social Research, No. 39, 1972, pp. 179–201

Cohen, E., 'A Phenomenology of Tourist Experiences', *Sociology*, No. 13, 1979, pp. 64–82

Conservation International Field Reports, *Ecotourism*, http://www.conservation. org/WEB/FIELDACT/C-C_PROG/ECON/ECOTOURHTM, accessed on 23/11/99

Cooper, C., J. Fletcher, D. Gilbert and S. Wanhill, *Tourism: Principles and Practice*, 1st edn, London, Pitman, 1993

Corrigan, P., *The Sociology of Consumption*, London, Sage, 1997

Craik, J., 'Are There Cultural Limits to Tourism?', *Journal of Sustainable Tourism*, Vol. 3, No. 2, 1995, pp. 87–98

Croall, J., *Preserve or Destroy: Tourism and the Environment*, London, Calouste Gulbenkian Foundation, 1995

De Kadt, E. (ed.), *Tourism – Passport to Development? Perspectives on the Social and Cultural Effects of Tourism in Developing Countries*, New York, Oxford University Press, 1979

Department for International Development, *Changing the Nature of Tourism: Developing an Agenda for Action*, London, Environmental Policy Department, DfID, undated

EIU, *Tourism and Developing Countries*, Travel and Tourism Analyst Occasional Studies, No. 6, 1989

Emery, F., 'Alternative Futures in Tourism', *International Journal of Tourism Management*, March, 1981, pp. 241–255

Featherstone, M., *Consumer Culture and Postmodernism*, London, Sage, 1991

Feifer, M., *Going Places: the Ways of the Tourist from Imperial Rome to the Present Day*, London, Macmillan, 1985

Fennell, D., *Ecotourism: an Introduction*, London, Routledge, 1997

Fukuyama, F., *The End of History and the Last Man*, London, Penguin, 1992

Furedi, F., *The Culture of Fear: Risk Taking and the Morality of Low Expectation*, London, Cassell, 1997

Fussell, P., *Abroad: British Literary Traveling Between the Wars*, Oxford, Oxford University Press, 1980

Ghazi, P., and J. Jones, *Downshifting: a Guide to Happier, Simpler Living*, London, Coronet, 1997

Giddens, A., *Beyond Left and Right: the Future of Radical Politics*, London, Polity Press, 1994

Goeldner, C., J. Brent Richie and R. McIntosh, *Tourism: Principles, Practices and Philosophies*, New York, Wiley, 2000

Goffman, I., *The Presentation of Self in Everyday Life*, Harmondsworth, Penguin, 1959

Goodman, D., and M. Redclift, *Environment and Development in Latin America: the Politics of Sustainability*, Manchester, Manchester University Press, 1991

Goodwin, H., 'Tourism and Natural Heritage, a Symbiotic Relationship?', in M. Robinson, P. Long, N. Evans, R. Sharpley and J. Swarbrooke (eds), *Environmental Management and Pathways to Sustainable Development*, Sunderland, Business Education Publishers Limited, 2000

Hall, C., and A. Lew, *Sustainable Tourism: a Geographical Perspective*, Harlow, Longman, 1998

Harrison, D., *Tourism and the Less Developed Countries*, London, Belhaven Press, 1992

Heartfield, J., *Need and Desire in the Post Material Economy*, Sheffield, Sheffield Hallam University Press, 2000

Hertz, N., *The Silent Takeover: Global Capitalism, and the Death of Democracy*, London, William Heinemann, 2001

Hewison, R., *The Heritage Industry*, London, Methuen, 1987

Hitchcock, M., V. King and M. Parnwell (eds), *Tourism in South-East Asia*, London, Routledge, 1992

Honey, M., *Ecotourism and Sustainable Development: Who Owns Paradise?*, Washington DC, Island Press, 1999

House, S., and K. Wood, *The Good Tourist*, London, Mandarin, 1991

Inskeep, E. (ed.), *National and Regional Tourism Planning*, London, WTO/ Routledge, 1994

Jenkins, T., D. Birkett, J. Butcher, P. Goldstein, H. Goodwin and K. Leech, *Ethical Tourism: Who Benefits?*, Debating Matters Series, Hodder & Stoughton/Institue of Ideas, London, 2002

Jones, J., 'Can Tourism Be Green', *Green World*, March, 1996

Kinnaird, V. and D. Hall (eds), *Tourism: a Gender Analysis*, Chichester, Wiley, 1994

Koshar, R., *German Travel Cultures*, Oxford, Berg, 2000

Krippendorf, J., *The Holiday Makers: Understanding the Impacts of Leisure and Travel*, Oxford, Heinemann, 1987

Lanfant, M., J. Allcock and E. Bruner (eds), *International Tourism: Identity and Change*, London, Sage, 1998

Lencek, L., and G. Bosker, *The Beach: the History of Paradise on Earth*, London, Pimlico, 1998

Lévi-Strauss, C., *Triste Tropiques*, Harmondsworth, Penguin, 1955

Lévi-Strauss, C., *The Naked Man*, translated by J. and D. Weightman, London, Harper Row, 1981 (originally published in 1971)

Lindberg, K., and D.E. Hawkins (eds), *Ecotourism for Planners and Managers*, North Bennington, VT, Ecotourism Society, 1993

Lodge, D., *Paradise News*, London, Penguin, 1991

Lury, C., *Consumer Culture*, London, Polity, 1996

MacCannell, D., *The Tourist: a New Theory of the Leisure Class*, London, Macmillan, 1976

MacCannell, D., *Empty Meeting Grounds: the Tourist Papers*, London, Routledge, 1992

MacMicheal, P., *Development and Social Change*, London, Sage, 1996

MacNaughten, P., and J. Urry, *Contested Natures*, London, Sage, 1998

Malik, K., *The Meaning of Race*, London, Macmillan, 1996

Malone, J., and N. Lewis, introduction in V.I. Lenin, *Imperialism: the Highest Stage of Capitalism*, London, Pluto, 1996

Mann, M., *The Community Tourism Guide: Exciting Holidays for Responsible Travellers*, London, Earthscan, 2000

Marcuse, H., *One Dimensional Man*, London, Routledge, 1999 (first published in 1964)

Mason, P., and M. Mowforth, *Codes of Conduct in Tourism*, Occasional Papers, University of Plymouth, 1995

McCormick, J., *The Global Environmental Movement*, Chichester, Wiley, 1995

McLaren, D., *Rethinking Tourism and Ecotravel: the Paving of Paradise and What You Can Do to Stop It*, West Hartford, CT, Kumarian Press, 1998

McNaught, T., 'Mass Tourism and the Dilemmas of Modernisation in Pacific Island Communities', *Annals of Tourism Research*, 9, 1982, pp. 359–381

Molstad, A., K. Lindberg, D. Hawkins and W. Jamieson, *Sustainable Tourism and Cultural Heritage: a Review of Development Assistance and its Potential to Promote Sustainability*, report for Norwegian World Heritage Organisation, 1999

Mowforth, M., and I. Munt, *Tourism and Sustainability: New Tourism in the Third World*, London, Routledge, 1998

Munt, I., 'Ecotourism or Egotourism?', *Race and Class*, Vol. 36, No. 1, July–Dec., 1994, pp. 49–60

Myerscough, J., 'The Recent History of the Use of Leisure Time', in I. Appleton, *Leisure, Research and Policy*, Edinburgh, Scottish Academic Press, 1974

Nash, D., 'Tourism as a Form of Imperialism', in V. Smith (ed.), *Hosts and Guests: the Anthropology of Tourism*, Philadelphia, University of Philadelphia Press, 1989

Nash, D., *Anthropology of Tourism*, Oxford, Pergamon, 1996

Nash, D., and V. Smith, 'Anthropology and Tourism', *Annals of Tourism Research*, Vol. 18, No. 1, 1999, pp. 12–25

Neale, G., *The Green Travel Guide*, London, Earthscan, 1998

Neale, G., *The Green Travel Guide*, 2nd edn, London, Earthscan, 1999

Omotayo Brown, D., 'Debt-funded Environmental Swaps in Africa: Vehicles for Tourism Development?', *Journal of Sustainable Tourism*, Vol. 6, No. 1, 1998, pp. 69–79

Omotayo Brown, D., 'In Search of an Appropriate Form of Tourism for Africa: Lessons From the Past and Suggestions For the Future', *Tourism Management*, Vol. 19, No. 3, 1998, pp. 237–245

Orams, M.B., 'Towards a More Desirable Form of Ecotourism', *Tourism Management*, Vol. 16, No. 1, 1995, pp. 3–8

Ousby, I., *The Englishman's England: Taste, Travel and the Rise of Tourism*, Cambridge, Cambridge University Press, 1990

Page, S., and R. Dowling, *Ecotourism*, London, Prentice Hall, 2002

Pattullo, P., *Last Resorts: the Cost of Tourism in the Caribbean*, London, Cassell, 1996

Pearce, D., *Tourist Development*, Harlow, Longman, 1989

Pepper, D., *Modern Environmentalism*, London, Routledge, 1996

Plogg, S., 'Why Destination Areas Rise and Fall in Popularity', in E. Kelly (ed.), *Domestic and International Tourism*, Wellesley, MA, Institute of Certified Travel Agents, 1997

Poon, A., 'Competitive Strategies for a "New" Tourism', in C. Cooper (ed.), *Progress in Tourism, Recreation and Hospitality Management*, Vol. 01London, Belhaven, 1989

Poon, A., *Tourism, Technology and Competitive Strategy*, Wallingford, CABI, 1993

Potter, R., T. Binns, J. Elliott and D. Smith, *Geographies of Development*, London, Prentice Hall, 1999

Price, M. (ed.), *People and Tourism in Fragile Environments*, Chichester, Wiley, 1996

Pryer, M., 'The Traveller as Destination Pioneer', *Progress in Tourism and Hospitality Research*, Vol. 3, 1997, pp. 225–237

Redclift, M., *Sustainable Development: Exploring the Contradictions*, London, Routledge

Richards, G., *Developing and Marketing Crafts Tourism*, Tilburg, ATLAS, 1998

Robins, K., 'Global Times: What in the World's Going On?', in P. du Gay (ed.), *Production of Culture/Cultures of Production*, London, Sage/Open University, 1997

Rojek, C., *Decentring Leisure: Rethinking Leisure Theory*, London, Sage, 1995

Ryan, C., *Recreational Tourism: a Social Science Perspective*, London, Routledge, 1991

Selwyn, T. (ed.), *The Tourist Image*, Chichester, Wiley, 1996

Sharpley, R., *Tourism, Tourists and Society*, 2nd edn, Huntingdon, Elm, 2000

Sharpley, R., 'In Defence of (Mass) Tourism', in M. Robinson, P. Long, N. Evans, R. Sharpley and J. Swarbrooke (eds), *Reflections on International Tourism: Environmental Management and Pathways to Sustainable Tourism*, Sunderland, Business Education Publishers Limited, 2001

Smith, V.L., *Tourism Alternatives*, London, Wiley, 1992

Smith, V.L., and W.R. Eadington (eds), *Tourism Alternatives: Potentials and Problems in the Development of Tourism*, Philadelphia, University of Pennsylvania Press, 1992

Srisang, K., 'Third World Tourism: the New Colonialism', *In Focus*, No. 4, 1992, pp. 2–3

Stabler, M. (ed.), *Sustainable Tourism: From Principles to Practice*, Wallingford, CABI, 1997

Sutherland, A., *The Making of Belize: Globalisation in the Margins*, Bergin and Garvey, Westpoint, 1998

Theobald, W., *Global Tourism: the Next Decade*, Oxford, Butterworth Heinemann, 1994

Therborn, G., *European Modernity and Beyond: the Trajectory of European Societies 1945–2000*, London, Sage, 1996

Turner, L., and J. Ash, *The Golden Hordes: International Tourism and the Pleasure Periphery*, London, Constable, 1975

UNEP, *Industry and Environment*, United Nations Environment Programme, Vol. 24, No. 3–4, July–Dec., 2001

United Nations, *Report to the United Conference on Environment and Development, Rio de Janiero*, 3–14 June (Vol. 11), United Nations, New York, 1992

Urry, J., *The Tourist Gaze: Leisure and Travel in Contemporary Society*, London, Sage, 1990

Urry, J., 'The Tourist Gaze and the Environment', *Theory, Culture and Society*, Vol. 19, 1990, pp. 1–26

US Agency for International Development, *Win-Win Approaches to Development and the Environment: Ecotourism and Biodiversity Conservation*, Centre for Development Information and Evaluation, USAID, Washington DC, July, 1996

Wahab, S., and J. Pigram, *Tourism Development and Growth*, London, Routledge, 1997

Wearing, S., and J. Neil, *Ecotourism: Impacts, Potentials and Possibilities*, London, Butterworth-Heinemann, 1999

Weaver, D.B., *Ecotourism in the Less Developed Countries*, Oxon, CAB International, 1998

Wheeler, B., 'Alternative Tourism – a Deceptive Ploy', in C.P. Cooper and A. Lockwood (eds), *Progress in Tourism, Recreation and Hospitality Management*, London, Belhaven, 1992, pp. 140–145

Wheeller, B., 'Eco or Ego Tourism: New Wave Tourism – a Short Critique', *Insights*, English Tourist Board, May, 1992

Williams, R., *Culture and Society*, Harmondsworth, Penguin, 1961

Williams, R., *Keywords*, London, Fontana, 1988

Withey, L., *Grand Tours and Cooks Tours: a History of Leisure Travel 1750–1915*, Aurum, London, 1997

Wood, R.E., 'Tourism, culture and the sociology of development', in M. Hitchcock *et al.* (eds), *Tourism in South-East Asia*, London, Routledge, 1992

World Tourism Organisation, *Global Tourism Forecasts*, WTO, Madrid, 1995

Wright, P.A., 'Sustainable Ecotourism: Balancing Economic, Environmental and Social Goals with an Ethical Framework', *Journal of Tourism Studies*, Vol. 4, No. 2, December 1993

WTO, *The British Ecotourism Market, special report no. 11*, World Tourism Organisation, Madrid, 2001

WTTC, WTO and Earth Council, *Agenda 21 for the Travel and Tourism Industry: Towards Environmentally Sustainable Development*, London, World Travel and Tourism Council, 1995

Index